The Story of Yellowstone

An Illustrated Natural History
From the Big Bang to the Big Burn and Beyond

Mike O'Connell

Lodgepole
Press

ISBN: 979-8-9858853-6-1 (hardcover)
ISBN: 979-8-9858853-9-2 (paperback)
ISBN: 979-8-9858853-3-0 (ebook)

Front cover image by Douglas Henderson

First hardcover edition, 2022

www.mike-oconnell.com

For Maeve

Contents

CHAPTER 1

The Big Bang, Earth's Fiery Birth and the Origins of Life

From blackness, from near nothingness, from a dense and hellish speck, our universe sprang to life. In that sliver of a second nearly 14 billion years ago, the subatomic building blocks of everything that would ever be raced outward. Today, these particles form the Beartooth Mountains, the Yellowstone River, cutthroat trout, lodgepole pine, wolves, elk and everything else in existence today in Yellowstone National Park and beyond.

As the rapidly expanding universe cooled, hydrogen and helium formed. Over time, hydrogen atoms joined together and billowed into enormous clouds in the dark infant universe. Gravity squeezed these clouds into dense cores millions of degrees hot, and nuclear reactions ignited these masses into fireballs, creating the first stars.

As they burned through their energy, the largest of the stars created a few dozen more elements, including carbon, oxygen, nitrogen, phosphorus and sulfur—necessary ingredients for the rise of life on Earth billions of years later.

When large stars burn through their energy after millions or billions of years, a core of iron remains, and nuclear fusion no longer takes place. The immense energy that radiated outward ceases. Gravity then pulls the stars in on themselves, and they explode as supernovas.

The intensity of these explosions creates all the other natural elements and sends them hurtling into space along with the lighter elements that formed earlier. Some bond together to make life-supporting molecules, including water, carbon dioxide and minerals.

Trillions of stars and planets in billions of galaxies have whirled to life from recycled gas and stardust. More than 4.5 billion years ago, some nine billion years after the birth of the universe, an exploding star jolted a cloud of this matter inside a spiral arm of the Milky Way Galaxy. Over a million years, this vast cloud flattened into a disk and spun faster and faster as gravity pulled the particles toward the center, where our hungry newborn sun consumed most of the dust and gas. Fusion from the massive stores of hydrogen ignited the sun, shining light and warmth on the remaining materials circling it.

Some of this matter sped farther away from the new star and

snowballed into the gaseous planets of Jupiter, Saturn, Uranus and Neptune. Closer to the sun, the coarser rocky material raced around and clumped together in collision after collision, creating Mercury, Venus, Earth and Mars, as well as asteroids and comets.

Every impact from ever-larger bodies released tremendous energy—so much so that the interiors of the four closest planets blazed as hot as the sun's surface and their exteriors melted into spheres of hot rock.

About 50 million years after Earth formed, Theia, a Mars-sized planet, slammed into her. The impact vaporized part of Earth and much of Theia. The rest of Theia's rocky remains plunged into Earth's molten interior.

Much of the vaporized rock rained back to Earth as fiery pellets, striking an endless sea of glowing lava. What remained in orbit clumped together in ever greater globs of molten rock. Soon, one sphere, having gobbled up virtually all the rest, dominated Earth's horizon and glowed red from the intense heat of its lava.

As an infant, the moon hovered close to Earth, as close as 12,000 miles away or about twenty times closer than today. This young moon spun around Earth frequently. About every four days, it eclipsed the sun, and in turn, Earth's shadow darkened the moon at the same rate.

Orbiting so close to Earth, the moon pulled relentlessly on her ocean of lava. Towering waves of this molten rock cascaded across the globe and splattered back down onto the red-hot sea. Likewise, Earth tugged the moon's molten landscape, but in the opposite direction. With so much turbulence, neither surface hardened at first; but with each orbit, the moon spun a little farther from Earth, and in time, the tidal forces lessened, and black basalt crusts formed on both surfaces amid the coldness of space.

As this cooling period began, asteroids and comets frequently slammed into the moon and Earth, ripping open their black crusts and revealing the fiery flesh below, and smaller meteorites relentlessly pelted both spheres. But as time went on, our solar system's planets and moons absorbed most of the rocky debris orbiting the sun, and the surfaces of Earth and our moon stabilized.

Yet because of the great heat and pressure deep below their surfaces, the Earth and its moon gave way to soaring fountains of lava that erupted from time to time. These volcanos billowed steam, nitrogen and other gases into the blistering, oxygenless air; and cones built up over time as the lava oozed and cooled.

Above Earth's surface, lightning blazed and thunder roared from dark, heavy clouds. Rain poured onto the planet's crust and boiled away. In time, Earth's surface cooled enough to hold water. Some rain still evaporated, but some found its way underground through cracks while some flowed to lower areas, slowly filling vast, shallow seas.

Earth's great internal heat melted basalt below the seafloor, releasing iron and other heavy components, which sank towards her core. Granite, a lighter and more buoyant rock, formed slowly from the transformed basalt. Slabs of granite rose above the basalt and peaked above the seas, creating islands, then microcontinents and eventually continents. The continental plates, set in motion by the currents of heat from Earth's mantle, slowly drifted across the globe. The North American plate, the home of Yellowstone National Park, started forming about three billion years ago.

The moon's gravity tugged on Earth's seas, creating immense waves that slammed into the land. The steady motion of these waves, combined with rainfall and gusty winds, gradually eroded granite and other rock. Grains of clay and sand flowed downstream to lower areas. The sunlight warmed the planet's beaches, tidal pools and clay-rich river deltas, yet no life existed anywhere on Earth—no bacteria, no algae, no plants, no animals.

Far beyond the light of the sun, deep below the surface of the sea, black plumes of superheated water, carbon dioxide, hydrogen sulfide, ammonia and other materials erupted from volcanic vents amid frigid seawater. Chemical reactions inside the heated vents transformed some of the water and other molecules into amino acids, lipids, sugars and other components needed for life. Some of these new molecules stuck to the vents and used minerals in the rock for energy. As they did so, they attracted other molecules and began stringing together.

After a nearly infinite number of arrangements, some of these molecules began to copy themselves. Eventually, they encased themselves in protective membranes that allowed them to leave the security of the vents and venture into the ocean, where they further diversified and thrived throughout Earth's seas.

For nearly three billion years, microorganisms like these made up all life on Earth. Close descendants of these microbes still thrive today in Yellowstone's hot springs and in the volcanic vents deep below the surface of Yellowstone Lake.

Within the first billion years after life formed on Earth, cyanobacteria evolved to make their energy from the sun's rays rather than the chemical energy of minerals. Later generations of cyanobacteria began releasing oxygen as a waste product, slowly adding that gas to Earth's oceans and atmosphere and dooming vast species of other microbes that found oxygen deadly. Increasing amounts of oxygen, however, set the stage for larger lifeforms to eventually evolve.

By a billion years ago, photosynthesizing green algae had appeared and flourished along the coast of Rodinia, a vast supercontinent pieced together from most of Earth's landmasses. At that time, the North American Continent, including Yellowstone, straddled the equator and was surrounded by the other landmasses. An expansive ocean surrounded Rodinia, and imposing waves pummeled and flooded Rodinia's barren lowlands.

The Big Bang, Earth's Fiery Birth and the Origins of Life

For millions of years, algae and other photosynthesizing microbes absorbed vast stores of carbon dioxide from the atmosphere. With much less of this heat-trapping gas in the atmosphere, Earth began cooling and reached a tipping point.

Steady snows covered lower and lower latitudes beginning more than 700 million years ago, and sheets of ice eventually spread from the poles toward the equator, thickening to nearly a mile deep in places. Life in the oceans either adapted to "Snowball Earth" or died. Some life-forms survived in and around undersea hydrothermal vents while photosynthesizers hung on in patches of open sea.

Were it not for her atmosphere and her blazing internal heat, Earth's oceans might have frozen solid and nearly all life would have died out. However, plumes of volcanic gas and ash exploded through the ice sheets, eventually filling the atmosphere again with carbon dioxide and other gases. With so few organisms surviving to absorb the carbon dioxide, Earth began warming, and the ice that had covered sea and land for millions of years began to melt. The resulting darker surfaces absorbed more of the sun's warmth than the highly reflective ice and further warmed the planet.

One effect of Snowball Earth was that, just as Yellowstone would experience many millions of years later, glaciers scraped away much of the Earth's soil and obliterated the geologic record in many places, including craters from asteroid and meteorite strikes.

As Snowball Earth warmed, nutrient-rich minerals from

glacially crushed rock streamed into the oceans, providing food for the surviving algae. They flourished again in a more tropical world and released increasing amounts of oxygen into the sea and air.

Life changed dramatically following Earth's greatest ice age. Some organisms evolved to reproduce sexually, rather than to simply divide, vastly speeding up evolution and the great variety of life-forms. Organisms began to grow shells, armor and spikes for protection, and they developed eyes, feelers, mouths, jaws and teeth in an increasingly sophisticated predator-and-prey world.

Higher oxygen levels in the sea provided more energy for animals to grow larger and to move more swiftly. These animals explored new habitats, chased down prey and darted away from predators.

The higher level of atmospheric oxygen and the development of a protective ozone layer also made life possible on land by shielding Earth from harmful solar radiation. Earth's magnetic field also protects the planet by deflecting solar wind and preventing it from ripping apart ozone and water vapor, which has been the fate of much of Mars' atmosphere.

Green algae, the ancestors of modern plants, pioneered life on land nearly 500 million years ago, at a time when Yellowstone lay under the sea near the equator. These ground-hugging explorers crept into swampy, low areas along the coasts and eventually evolved into moss-like, leafless plants with shallow roots. Anchored by their roots, they drew water from the rocky ground as their stems and branches harnessed the sun's energy. As generation after generation of plants died, their decaying matter built up soil for more diverse vegetation, which began spreading around the globe.

Plants became sturdier over time; by 360 million years ago, they began to grow much taller, developed leaves and absorbed more energy from the sun. Soil and decaying matter grew deeper and more widespread. Huge ferns, horsetails, primitive trees and other leafy plants made up the first forests and turned Earth's surface from a rusty Mars-like color to green. Prototaxites, towering fingers of fungi growing to nearly 30 feet high, rose up from the soil as well.

At this time, the Yellowstone area sat in the tropics under a warm, shallow sea full of small-shelled creatures, including mollusks, crinoids, brachiopods and bryozoans. Fossils of these creatures can be found today in sedimentary rock in Yellowstone and the lands around it. These organisms secreted calcium carbonate to form their protective shells, and when they died, their bodies settled on the seafloor. Over millions of years, the shells compacted and formed layers of limestone up to 2,000 feet thick. Today in Mammoth Hot Springs, cold water from rain and snowmelt seeps down thousands of feet to this buried limestone, where it absorbs large amounts of calcium, carbon dioxide, sulfur and other materials. Pressurized at depth and heated by the magma below Yellowstone's surface, the water rises, emerges in Mammoth's hot springs and cools. Minerals filter out and form travertine, the rock that makes up the chalky white terraces there.

Before plants colonized land, the ancestors of spiders, scorpions, millipedes and other invertebrates ventured onto Earth's shores to flee predators and follow prey. After Earth began greening, visits became permanent and these creatures diversified rapidly, filling new ecological niches. They ate any available food, from vegetation and decaying plants to each other.

Fish similarly flirted with land to find food and to avoid being eaten. Some evolved limb-like fins, allowing them to crawl out of the sea for limited forays. In time, their fins became true limbs, their air bladders morphed into lungs and their bones strengthened. These first amphibians still depended on water to lay their soft eggs, but they could now move farther from the sea and thrive in new habitats where insects teemed.

By 330 million years ago, most of the continents came together again to form the supercontinent Pangaea. Around this time, shallow seas ebbed and flowed over the flat, mountainless Yellowstone region. Life abounded on Pangaea in the lush, swampy forests of tree ferns, conifers and tree-sized, Y-shaped club mosses called lycopods. Dragonflies with two-foot wingspans zipped over the steamy land, snagging other insects from the air. Their larvae in the waters below ate just about anything smaller than themselves, including fish and amphibians. Cockroaches and other insects rustled in the leaf litter and worms tunneled through the rich soil.

The first reptiles appeared on Pangaea about 315 million years ago. These small, agile and scaly-skinned animals evolved from amphibians and laid hard-shelled eggs on land, which hatched smaller versions of their parents rather than an intermediate-staged tadpole. Without having to depend on water for sustaining eggs, reptiles could venture more widely than their ancestors and soon came to dominate the land.

The Story of Yellowstone

CHAPTER 2

Dinosaurs Rule

About 252 million years ago, massive volcanic eruptions flooded more than a million square miles of ancient Siberia with lava, harkening back to the days of a molten young Earth. Enormous amounts of carbon dioxide, sulfur dioxide and other gases billowed into the sky. Over time, the massive amounts of carbon dioxide and other greenhouse gases in the atmosphere trapped the heat radiating from Earth, broiling the planet.

This led to warmer and more acidic oceans containing much less oxygen. Die-offs at the lowest rungs of the food chain rippled upward and ultimately caused the collapse of entire ecosystems. More than 90 percent of all aquatic and terrestrial species perished in this great mass extinction, the worst Earth has ever experienced.

As the Triassic period dawned following this massive die-off,

Earth's climate stabilized but remained hot and dry, especially inland from the coast of Pangaea, where vast deserts spanned the supercontinent. From the start of the Triassic until about 83 million years ago, shallow seas occasionally covered the Yellowstone area or lapped its edges, sometimes for millions of years at a time. More often than not, however, the area remained beyond the reach of the sea.

Smaller animals that survived the mass extinction rapidly diversified after Earth became hospitable again. This included the first dinosaurs, turtles, lizards, crocodiles and mammals. As time went on, coastal forests of conifers, ginkgoes and palm-like cycads spread nearly to the ice-free poles. Animals also expanded their ranges and grew ever larger.

During the Triassic, some land-dwelling reptiles that had evolved from sea-based amphibians returned to the ocean. This group included ancestors of the long-necked, surface-breathing plesiosaurs, which ranged in size from 5 to 50 feet long. Re-adapting to life in the ocean, their legs morphed into powerful paddles, which they used to chase down fish, squid and other prey, sometimes above a sunken Yellowstone landscape.

Ichthyosaurs, which were dolphin-like reptiles and a top ocean predator, hunted over the area as well, swiftly catching prey with their streamlined skulls and sharp teeth. The warm ocean also teemed with ammonites, mollusks, urchins and plankton. And by

220 million years ago, the reptilian pterosaurs fished and hunted from the skies.

By 200 million years ago, Pangaea began breaking apart. Volcanos erupted and earthquakes shook the land as the North American Continent broke away from the South American and African Continents. The extensive volcanic activity at this time quadrupled the amount of carbon dioxide in the atmosphere, warming the planet and acidifying the oceans. This triggered another great die-off when nearly half of all land and marine animal species went extinct.

Reptilian survivors of the die-off at the end of the Triassic and their descendants, including pterosaur, dinosaur and plesiosaur species, would continue to rule the air, land and sea for nearly 135 million more years. Mammals also survived the mass extinction at the end of the Triassic but remained small and lived in the shadows of the dinosaurs until the latter died out 65 million years ago.

During the Jurassic period, which spanned about 200 to 145 million years ago, Earth's climate grew humid and subtropical. Coniferous forests, ferns, horsetails and cycads spread into the old Triassic deserts. Lush vegetation supported abundant life and larger and larger dinosaurs, including the long-necked sauropods, the largest animals ever to walk the Earth.

Warm shallow seas occasionally covered Yellowstone Country during the middle of the Jurassic. Plesiosaurs, ichthyosaurs, crocodiles, sharks, turtles and rays hunted in these waters at those times. When the sea receded, wide forested coastal plains with large meandering rivers hosted a diverse variety of life.

Herds of slender *Diplodocus* moved through the forests, browsing on ferns and stretching their long necks into the trees, where they stripped leaves from the branches. These sauropods reached lengths of more than 100 feet and weighed up to 18 tons. When alarmed, they cracked their long, whip-like tails at supersonic speed. A strike from one of them could topple a predator and even kill it.

Allosaurus, a top hunter of the time, probably avoided attacking such large animals, and likely ambushed the young, old and weak, as well as scavenged carcasses. The powerful *Allosaurus*, an ancestor of *Tyrannosaurus rex*, grew to 30 feet long, weighed nearly a ton and sported sharp, 10-inch claws. It likely opened its massive jaws wide to slash at prey with its long, knife-like teeth.

Other dinosaurs in the Yellowstone area included *Apatosaurus*, another long-necked herbivore with a whip-like tail but a stockier build than *Diplodocus*. With their huge, rounded feet and weighing up to 30 tons, they could flatten predators, while their claws provided further protection.

Allosaurus found smaller prey in *Camptosaurus*, a duck-billed dinosaur. This bulky and common browser weighed a little less than *Allosaurus* and probably walked on four legs when foraging, but on two legs when traveling or fleeing danger.

Stegosaurus also kept an eye out for *Allosaurus* as it browsed on low-growing plants, such as ferns. It weighed up to two tons and grew to 30 feet long. Spikes up to three feet tall on its tail, which it held horizontal as it plodded through Jurassic forests, defended it from predators, while the pointed plates on its back were likely for display.

The planet's first birds evolved from small meat-eating dinosaurs by the late Jurassic, joining the pterosaurs in the air. The descendants of these flying dinosaurs live on in Yellowstone today, and some— such as ravens, magpies and eagles—probably behave similarly, raiding nests to feed on eggs and chicks or scavenging meat from carcasses.

No mass extinction marked the end of the Jurassic and the beginning of the Cretaceous, but by this time, many of the Jurassic dinosaurs that lived in the Yellowstone area had died out, including *Diplodocus, Apatosaurus, Camptosaurus, Stegosaurus* and *Allosaurus*. However, dinosaurs here and around the world continued to diversify and thrive.

Pangaea continued to fracture into smaller continental pieces during the Cretaceous, which spanned 145 to 65 million years ago; by the end of the period, the continents had migrated roughly to their present positions.

A warm and shallow inland sea, which occasionally stretched

from the present-day Gulf of Mexico to the Arctic, periodically flooded Yellowstone Country during the Cretaceous period. When the sea was present, plesiosaurs, ichthyosaurs, turtles, crocodiles, sharks and rays swam above Yellowstone Country as they sometimes had during the Jurassic. Fossil fragments of these creatures lie buried in sedimentary rock from the ancient seafloor in places like Yellowstone's Mount Everts, the prominent ridge to the east of Mammoth Hot Springs. In a one-day survey of Mount Everts several decades ago, paleontologists discovered the skeleton of a plesiosaur and fragments of a turtle shell and dinosaur egg from the Cretaceous period.

New species of pterosaurs and birds flew over the area while small mammals scurried along the forest floor whenever the seas receded. By 110 million years ago, flowering plants had evolved and spread around the globe with the help of bees, ants, beetles and other pollinators.

Around the same time, the plant-eating dinosaurs *Titanosaurus* and *Tenontosaurus* walked the ancient Yellowstone landscape. Titanosaur, a long-necked dinosaur, reached lengths of up to 45 feet, about half the size of *Diplodocus* from 40 million years earlier. And at a maximum of six tons, it weighed about a third as much as its predecessor.

Tenontosaurus primarily walked on two legs, but also dropped to all four to forage. It grew to 25 feet long, weighed a ton and probably lived in groups, giving it protection from one of the fiercest dinosaurs to ever live—*Deinonychus*.

An 11-foot-long relative of velociraptors, *Deinonychus* had long arms and fingers with razor-like claws. These 160-pound predators probably hunted in packs and jumped on prey, seizing them with their claws and slashing them open with sickle-like talons on their hind feet.

Sauropelta, an armored dinosaur, also lived in the region at the same time as *Deinonychus*, *Titanosaurus* and *Tenontosaurus*. A four-legged herbivore, it grew to 18 feet long and weighed three tons or more. Its armor and large neck spikes protected it from predators.

During the latter part of the Cretaceous, the Rocky Mountains began forming to the west of Yellowstone Country, and by 83 million years ago, the sea drained away from the Yellowstone area for good as the rising landscape pushed the waters to the east.

Tyrannosaurus rex arose in the shadow of the Rockies and thrived in the ancient American West. *T. rex* stood up to 40 feet tall and weighed as much as 12,000 pounds. Its nearly foot-long serrated teeth filled its powerful bone-crunching jaws, and it likely scavenged and stole carcasses from smaller predators at least as often as it brought down its own prey. In Montana, Wyoming and the Yellowstone area in general, its diet included *Triceratops* and *Edmontosaurus*, a duck-billed dinosaur.

Edmontosaurus and *Triceratops* may have traveled in herds of thousands, not unlike the large herds of bison that did so for thousands of years in western North America. Individuals from both groups weighed as much as five tons, though at nearly four feet long, *Edmontosaurus* grew about a third longer than the stockier *Triceratops*, whose three horns and large neck shield might have been used for display rather than defense.

Ankylosaurus, another herbivorous dinosaur from this time, weighed nearly as much as *Edmontosaurus* and *Triceratops*, though it was shorter and stockier at 16 feet long and more than six feet wide. Its armor plating and huge, bony, club-like tail protected it from predators. These protections, however, could not save *Ankylosaurus* and other dinosaurs from forces from the skies above and from deep within Earth.

CHAPTER 3

The Age of Mammals

In the blink of an eye 65 million years ago, a six-mile-wide asteroid, streaking at 40,000 miles an hour, slammed into Mexico's Yucatán Peninsula. It tore deep into the shallow seabed, heating the strike zone to thousands of degrees Fahrenheit and blowing out a 110-mile-wide impact crater.

As it smashed into Earth, the trillion-ton asteroid vaporized into a fiery plume, shooting billions of tons of molten rock and dust into the atmosphere and across sea and land.

The blast destroyed everything in its path within a six-hundred-mile radius. Massive earthquakes rocked the region, volcanos erupted, and shockwaves traveled thousands of miles around the globe. Tsunamis a thousand feet high or more wiped out nearly all life within coastal and low-lying areas for thousands of miles.

As hot rock rained back to Earth, the atmosphere heated up and ocean surfaces steamed. Plants and forests smoldered and burned throughout North and Central America, and the skies darkened with smoke and dust that blocked much of the sunlight. As a result, Earth cooled, and for many months, the planet remained dim and chilly. Photosynthesis stalled, and plant and animal life suffered greatly.

Countless dinosaurs, pterosaurs and plesiosaurs died upon the asteroid's impact, during its immediate aftermath and through the prolonged winter that followed. It took thousands of years for them to completely die out, but when they did, they joined as much as 75 percent of Earth's species that also perished after the asteroid strike.

Earth, however, might already have been nearing a tipping point. The asteroid might simply have struck the final, staggering blow.

Extensive lava flows in ancient India's Deccan Traps had begun flooding an area the size of California hundreds of thousands of years before the asteroid rocked Earth. The Yucatán blast likely ramped up the activity in these volcanic fields. Immense levels of carbon dioxide and other gases released from both the Deccan volcanos and the asteroid's collision filled the atmosphere. These gases trapped heat, and the planet warmed greatly after her skies cleared from the impact's smoke and dust. In time, the climate moderated as the greenhouse gases dissipated. The plants and animals that had clung on in the bleakness sprang back.

The worldwide vacuum left by the dinosaurs allowed mammals

to step out of the shadows, to diversify and to get bigger. Up until then, they had typically foraged and hunted in the forest, usually at night. Their burrows had protected them from the chaotic climate swings, and similarly, deeper waters sheltered many fish, turtle, crocodile and alligator species. As with mammals, the smaller the reptile, amphibian or bird, the more likely it was to survive and diversify after the die-off.

As forests and plants rejuvenated throughout North America, which bore the brunt of the asteroid's impact, mammals moved back in. Some of these colonizers emerged from protected areas or migrated from the Arctic, which was much warmer then, ice-free and thousands of miles from the impact site. Others crossed into North America from Asia whenever land connected the two continents, a common occurrence going back nearly 100 million years. Rising sea levels, however, flooded the land bridge between the continents by 10,000 years ago.

Although mammals generally remained small during the dinosaurs' reign, they had been evolving for more than 150 million years during that time. Some of these laid external eggs, like today's duck-billed platypuses and echidnas. Others evolved to give birth to live young, including marsupials, like today's kangaroos and koalas, and placentals. The latter two groups comprise the vast majority of mammals alive today. The young of both marsupials and placentals begin life inside their mothers, where they are nourished by a placenta, but underdeveloped marsupials emerge earlier and continue to develop outside of the womb, usually in a pouch.

Marsupials thrived in North America alongside dinosaurs. Some of the oldest known marsupial fossils come from Montana and North Dakota. After the dinosaurs died out, placental mammals became more common than marsupials, and egg-laying mammals grew exceedingly rare.

Within half a million years after the asteroid struck, several species of small-hoofed mammals lived in the Yellowstone area. As they continued to diversify, many grew larger and began to fill the dinosaurs' niches in forests and along streams.

Birds also rapidly diversified and spread throughout the world. And in time, some land mammals adapted to ocean life and evolved into whales, porpoises and dolphins.

By 50 million years ago, uintatheres, the largest animals of the time, arose in Asia and North America and expanded and diversified throughout both continents, including Yellowstone. Some of these herbivores resembled and grew as large as rhinoceroses. Their tusks probably helped them to defend themselves and to gather and eat aquatic vegetation.

Titanotheres were another rhinoceros-like animal that lived in the Yellowstone area until about 34 million years ago. In 1999, a titanothere tooth and mandible were found on Mount Hornaday near Yellowstone's Lamar Valley. When they lived, the climate was warmer and more humid, and the park's lower elevations would have resembled the tropical forests found in Central America today.

Ancestors of even-toed animals, such as deer, pronghorn antelope and camels, and odd-toed animals, including horses, rhinos and tapirs, lived alongside the uintatheres and titanotheres in North America. *Hyaenodons*, which resembled hyenas but weren't related, hunted these and other hoofed animals in the area before dying out by 23 million years ago. They ranged in size from a foot to more than nine feet long.

Hesperocyon, the raccoon-like ancestors of wolves, coyotes and foxes, evolved in North America by 40 million years ago. They ambled around forests on short legs and ate small prey, fruits and seeds.

Coyote-sized, three-toed horses also ate fruits and seeds, as well as the leaves and plants that grew in the tropical forests. Their earliest ancestors were even smaller, but newer generations continued to grow larger and diversify over time.

After the climate dried and cooled by about 33 million years ago, grasslands began replacing forests throughout much of western North America, and forest-dwelling, browsing horses gradually gave way to grazing horses. Some, such as the three-foot-tall *Merychippus*, lived in large herds in the open grasslands 16 to 10 million years ago, including the Yellowstone area. Their teeth evolved to grind the abrasive grasses, and their middle toe evolved into a hoof, making them agile runners. Speedier horses with longer legs and single-hoofed feet evolved later in response to swift predators.

Horses flourished in North America for more than 50 million years before dying out by 10,000 years ago after the end of the last ice age. Later, the Spanish introduced modern horses to North America in the 1500s, and by the early 1700s, horses had reached the Yellowstone area.

Ancestors of rhinoceroses, peccaries, beavers, pocket gophers and other rodents had also crossed onto the continent by 33 million years ago, as well as species of the cud-chewing Ruminantia, which gave rise to pronghorn antelope, mountain goats, bighorn sheep, deer and others. The felid-like Nimravidae also reached North America by then. *Pogonodon,* a Nimravid that looked like a saber-toothed cat, stalked prey in the area. However, the true ancestors of lions, cougars, lynx, bobcats and other felids migrated from Eurasia millions of years later. The Nimravidae had died out worldwide before then.

The climate warmed again by 24 million years ago and remained relatively mild, though dry, for about 10 million years. About 17 million years ago, the elephant-like mastodons and four-tusked gomphotheres arrived in North America, becoming the largest species to inhabit the continent. The biggest animals until that point, the North American rhinos, coexisted with mastodons and gomphotheres for millions of years until dying out by 3 to 4 million years ago.

Teleoceras, a horned rhino that inhabited the Yellowstone area, looked more like a hippo, with its barrel-shaped chest and stumpy legs. These 13-foot-long rhinos lived close to ponds and streams and grazed on grasses. *Amebelodon*, a gomphothere that lived in the area about the same time, browsed rather than grazed, and likely used its slender, shovel-like lower tusks to strip leaves and bark from trees and to scrape up vegetation. A little smaller than modern African elephants, *Amebelodon's* ancestors lived on that continent before some migrated into Eurasia and then to North America.

By 19 million years ago, pronghorn antelope arose on the North America grasslands, and by five million years ago, at least 12 species lived there. The horns of some of these antelope resembled deer antlers, while others looked like small moose antlers. Another species' horns spiraled straight up above their heads while another one grew six horns.

Later pronghorn species evolved alongside the speedy American cheetahs, which were closely related to cougars rather than today's African cheetahs and died out by 10,000 years ago. Pronghorn developed large windpipes, hearts and lungs, enabling them to race from danger at nearly 60 miles an hour. They are the fastest land animal in North America today and the second fastest in the world behind African cheetahs.

By 3 million years ago, horses, wolves and foxes had also moved across the Bering Land Bridge into Eurasia, and some ventured farther to Africa, including the ancestors of zebras, Ethiopian wolves and fennec foxes.

Camel species migrated to Eurasia and Africa millions of years ago, as well, adapting to deserts and other marginal places. And they moved into South America, eventually giving rise to llamas, alpacas, vicuñas and guanacos.

All in all, more Eurasian species moved into North America than the reverse, but North American species of all kinds traveled much more readily into South America than southern species moved north. This began about 10 million years ago and became more common about 3 million years ago when the Panama land bridge formed, linking North and Central America with South America. Southward migrants included mastodons, bears, tapirs, horses, camels, canids, raccoons, peccaries, deer and saber-toothed cats. Some of the more successful species to move north from South America and eventually reach the Yellowstone area included ground sloths, porcupines, snakes and toads.

By about 2 million years ago, many of North America's horse, camel and pronghorn species had died out, joining the fate of its rhinos. This happened as the global climate continued to cool, ultimately leading to the ice ages.

By 1.7 million years ago, mammoths crossed into North America from Eurasia, joining the mastodons, which had been on the continent for 15 million years. Mastodons generally stayed closer to forests and browsed while mammoths, which evolved flatter molars, grazed on and crushed tough grasses and vegetation.

By 200,000 years ago, herds of long-horned bison also grazed the Yellowstone area's grasslands. They evolved from earlier bison that had migrated into North America from Eurasia. Long-horned bison weighed nearly twice as much as their relatives roaming Yellowstone National Park today. The largest bulls' horns spanned nearly seven feet, more than three times as wide as today's bison's more curved horns. Long-horned bison died out by 20,000 years ago near the peak of the most recent glaciation. Their smaller relatives eventually filled the void left by them.

Other ungulates that lived in the area by 100,000 years ago included white-tailed and mule deer, musk ox, caribou, moose and stag-moose, which grew larger than modern moose but resembled elk, with palmate antlers. These animals lived alongside at least three species of horse that adapted to the global cooling, as well as at least five species of pronghorn, elk-sized mountain deer, camels, ground sloths and black-bear-sized beavers.

A diverse array of carnivores stalked these herbivores in the forests, tundra and grasslands of Yellowstone Country. This included giant short-faced bears, the largest mammalian land carnivores ever, weighing up to 2,200 pounds and standing 12 feet tall on their hind feet. With long, slender legs, these "running bears" probably chased down some prey, but similar to grizzly bears today, they likely also stole carcasses from smaller predators, such as wolves and lions.

Grizzlies arrived in the area from Alaska and Canada within the past 15,000 years, while the more omnivorous black bears appeared more than a million years ago and spread throughout the continent.

The American lion, the largest cat ever, weighed up to 1,100 pounds and likely hunted small and large prey alike—from deer, horses and camels to bison and mammoths. To attack larger animals, the lions probably hunted in small groups as African lions do today.

The long-legged scimitar cat, another ice-age feline, grew as large as modern African lions and ran down young mammoths, mastodons and other herbivores, dragging prey to their rocky dens to feed themselves and their kits. Bites from their long, saw-like canine teeth bled prey to death and disemboweled it.

Smilodon, another saber-toothed cat with much longer canines and a powerfully built body, ambushed large prey and wrestled it to the ground. It likely also hunted larger animals cooperatively.

Dire wolves, another carnivore that hunted together during the Pleistocene, grew about as tall as today's longer-legged gray wolves, but at up to 150 pounds, they weighed about a third more. They probably also scavenged on carcasses. By 15,000 years ago, gray wolves hunted ungulates in the area as they do today. A third wolf, the Beringian wolf, fell between dire and gray wolves in size. This longer-snouted canid belonged to the same species as gray wolves but evolved separately and might have specialized in hunting migratory musk ox and bison.

As the ice disappeared from all but the highest points by 10,000 years ago, nearly 75 percent of all the large predators and prey also vanished from North America and the greater Yellowstone landscape, including short-faced bears, *Smilodon*, scimitar cats, American cheetahs, mastodons, mammoths, horses, camels, peccaries, mountain deer, stag-moose, sloths and all but one pronghorn species. Others died out within a few thousand years after this, such as dire and Beringian wolves, while a few found refuge in far-off lands, including caribou and musk ox in the high Arctic.

Climate change, notably a quick drop in global temperature about 13,000 years ago, might have contributed to the die-offs, but human migrants from Eurasia likely played the largest role. By 13,000 years ago, they had spread throughout North America, including the Yellowstone area. The oldest human remains found so far in the Americas date to then and belong to a young boy whose nomadic family buried him about 80 miles north of Yellowstone in the Shields Valley, near what's now Wilsall, Montana. The boy's family buried him along with more than 100 stone and bone hunting and animal-processing tools, presumedly for hunting in the afterlife.

Being buried with those tools underscored the importance of

hunting for the Clovis people living in North America back then and the advantage it gave them over their prey. Armed with spears, they might have overhunted mammoths, mastodons and other herbivores and outcompeted other carnivores. As their prey dwindled, so did the big cats, wolves and bears.

Large animals that survived the die-off, including bison, bighorn sheep, moose, grizzlies, black bears, pronghorn, cougars and gray wolves, adapted to the new hunters.

Some species, such as elk, evolved alongside humans in Eurasia, maybe giving them an advantage over other species unfamiliar with people. When new foraging grounds opened up in North America following the reduction or extinction of other species, elk migrated into these areas.

As human populations have steadily increased throughout the world following the last ice age, wild landscapes have shrunk considerably. Correspondingly, wildlife has also been getting smaller in size and in numbers over time. And some wildlife species have shifted from being active during the day to moving about at night to avoid people.

This adaptability has served mammals well for more than 200 million years; and in Yellowstone Country, perhaps unlike anywhere else in the world, they've adapted to a wide range of other challenges besides people—from volcanos to glaciers to fire.

CHAPTER 4

Mountains Rising

Beginning about 170 million years ago, during the Jurassic and long before mammals came to dominate Earth, flatlands in present-day Washington, Oregon and California began rising as the westward-advancing North American continental plate rode over denser oceanic plate. The sinking seafloor heated as it met hot rock deep below and rose as magma that formed chains of volcanos inland from the shore.

Nearly 90 million years later, mountain building reached the western edge of Yellowstone Country. Mountains had existed in the area more than three billion years before that and more recently but had eroded away. The rolling Appalachian Mountains in the eastern United States, which once rivaled the Himalayas in elevation, have been experiencing a similar fate of erosion for hundreds of millions of years.

Sedimentary rock, formed from old seafloor and mud flats, thrusted up in layers, overlapping like fish scales from west to east, and extending from Canada to Mexico. As the flatlands rose in Idaho, Montana and Wyoming, the inland sea shifted eastward and out of Yellowstone once and for all.

About 60 million years ago, massive blocks of nearly 3-billion-year-old granite and gneiss bedrock began rising beyond the northeast corner of Yellowstone, forming the Beartooth Plateau and the Beartooth Mountains. The Beartooth uplift rose along much deeper faults than the earlier mountain-building to the west. In places, it pushed up ancient layers of seafloor more than two miles above sea level.

Some of the rock in the Beartooths encases four-billion-year-old zircon crystals, tough minerals that are born in molten rock, erode out from their parent stone and later mix into sedimentary and metamorphic rock, including gneiss and schist.

The Beartooth Plateau runs for more than 70 miles, from southeast of Livingston, Montana, to south of Red Lodge, Montana. More than 20 peaks in the Beartooths top 12,000 feet, including Granite Peak, the highest point in Montana at nearly 13,000 feet.

Over time, the softer sedimentary layers on the Beartooth bedrock wore away, exposing the underlying granite, gneiss and schist. Tens of millions of years later, glaciers sculpted the Beartooths and carried boulders down into Yellowstone, dropping them in the Lamar Valley and other areas.

About 10 million years after the Beartooths began forming, volcanos started erupting throughout Yellowstone Country. The earliest volcanos from this period, the Absaroka Volcanic Supergroup era, created the Washburn Group. This group's two parallel mountain chains run from roughly northwest of present-day Mammoth Hot Springs southeast to beyond Yellowstone's Canyon Village, and from southeast of Emigrant, Montana, to the southeast beyond Cooke City, Montana. These mountain chains include the Washburn Range in the north-central area of the park and parts of the Gallatin Range in the northwest of Yellowstone and extending northward outside of the park.

Along with the Washburn Group, the Absaroka Volcanic Supergroup includes mountain chains formed later by the Sunlight and Thorofare Group's volcanos inside and outside of the park. The Sunlight Group include mountains mostly east of the park while the Thorofare Group's mountains run mainly southeast of Yellowstone.

Volcanic rock from the Thorofare Group, the youngest, rests on top of the older two in places, and in other areas, Sunlight Group rock overlays Washburn-era rock.

Each group's volcanos built up slowly following eruptions when lava, rock and ash spilled from vents and settled. Like modern-day stratovolcanoes in the Cascades and Andes, these cones largely formed from andesite, a mix of basalt and silica-rich rhyolite.

Eruptions ranged from relatively quiet ones with lava flowing downslope to more explosive ones, which shot out ash and rock bombs and sent fast-moving tides of ash, lava and rock into the valleys. The larger and heavier the rock, the more likely it settled closer to the cones. Lighter material, including ash and pebbles, traveled the farthest.

The cones of the Absaroka Volcanic Supergroup reached elevations of more than 10,000 feet and dominated the landscape. Dense forests of pine, fir and spruce grew in the higher altitudes while tropical trees and plants grew thickly in the valleys and lower elevations, at a time when Yellowstone Country and the world were much warmer. Trees growing in the region included redwoods (the most abundant type), sequoia and other conifers; walnut, oak, maple, sycamore, chestnut, elm, ash, hickory, dogwood, laurel, alder and willow; and subtropical and tropical species, including breadfruit, persimmon, avocado, fig, magnolia and cinnamon. In the drier and more temperate climate today, fewer than a dozen tree species grow in the area. Nearly all these present-day species, except aspen and cottonwood, are conifers.

During eruptions, heat from escaping lava melted snow and ice high on the mountain slopes and mixed with ash and other debris. The resulting mudflows, or "lahars," thickened as they accumulated more ash, as well as sand, gravel and rock from the stream beds they flowed into and scoured. They flowed powerfully down the steep-sloped volcanos, toppling large sections of forest and carrying boulders, trees and stumps with them.

The lahars could run for miles and fanned out hundreds of yards as they slowed in the valleys. They buried lower-elevation trees where they grew and deposited higher-elevation ones, sometimes upright. Needles, cones, leaves, plants and pollen from alpine to tropical zones likewise mixed in the valleys. Over time, lahars cemented into conglomerate—a mix of compressed mud, ash and pebble- to boulder-sized rocks—that can be seen throughout the Yellowstone region today, including the Mount Washburn area and the Lamar and Gallatin Valleys.

After being buried and protected from chewing insects and the corrosive effects of oxygen, trees sucked up silica-rich water from dissolved volcanic ash. This water seeped into every cell of the wood, and silica replaced the tissue over time. The silica molecules then formed quartz crystals and fossilized into stone—stone that often preserved the fine detail of tree rings, cells, knots and sometimes bark, especially in the fossilized forests of the Lamar and Gallatin Valleys. Some needles, cones and leaves also petrified.

After a lahar clears a swath of forest, life eventually returns to the lunar-like landscape. In sheltered and moist areas, including the deep hoof prints of animals crossing the barren ground, seeds blow in from untouched areas, take hold and sprout. In other places, plants spring up from broken pieces of root. Spiders and insects crawl or fly in as tiny trailblazers amidst the new plants and saplings. In turn, hungry birds reappear. Carnivorous animals likewise follow the returning herbivores. And over time, plant pioneers give way to successive ones and mature forests thrive again.

Similar to Mount St. Helens in the Cascades, the supergroup volcanos lay dormant for hundreds to thousands of years between eruptions, allowing forests to regrow.

Over millions of years, countless lahars raced down the volcanos and buried forests, sometimes on top of previously entombed trees. On Specimen Ridge, in the northeast of Yellowstone, 27 layers of forest lie one on top of the other. In some places, standing trees from one layer poke up into an overlying one.

By 43 million years ago, the volcanos of the supergroup went silent. Lahars occasionally rushed down mountain slopes and leveled trees after heavy rains, but no new ash or lava spewed from the vents.

An area about two-and-a-half times the size of Yellowstone National Park lay covered in conglomerate—in places to nearly a mile deep. Only the highest peaks in the park and surrounding area poked through these millions of acres of volcanic debris, covered thickly by forest and zigzagged here and there by streams. These streams carved deep valleys over time into the soft, rocky ground, and the mountains of conglomerate steadily wore down for millions of years—so much so that today's mountain tops, including Druid and Barronette Peaks, reach only as high as ancient valleys.

Some of the hard-rock interiors of the volcanos still stand, including Sunlight Peak, east of Yellowstone; Meldrum Mountain, Bunsen Peak and Black Butte in the Gallatin Range; Mount Washburn of the Washburn Group; Lake Butte, northeast of Yellowstone Lake; and Colter Peak in southeast Yellowstone.

By and large—except for the ongoing rise of the rugged Teton Range within the past 10 million years—Yellowstone Country sat relatively quiet for more than 40 million years until a new era of volcanism arrived and radically changed the land.

The Story of Yellowstone

CHAPTER 5

The Yellowstone Supervolcano

One day just over two million years ago, Yellowstone shook violently, spooking wildlife for miles around, from mastodons browsing in forests to cheetahs hunting pronghorn in grasslands. Following the earthquakes, an enormous cloud of billowing ash, rock and gas erupted from the ground with a deafening roar and shot more than a dozen miles skyward at supersonic speed.

Hellish waves of rock, gas and ash raced from the volcano and across the land at more than 100 miles an hour. These pyroclastic flows incinerated forests, animals and birds, and vaporized streams and all living things in them. Whatever life escaped the 1,500-degree-Fahrenheit flows could not escape the rock bombs and heavy ash falling from the collapsing column above.

Later, two more colossal explosions rocked the area from

different vents, shooting more ash and rock into the atmosphere and sending more lava racing across the land.

For many miles around the blast zones, the smoldering, ashy land took on a lunar appearance, including the 60-by-40-mile crater ripped open by the three explosions that had obliterated the mountains and valleys and all living creatures.

Ash from these eruptions settled nearly a thousand miles to the east, to the Mississippi River, and reached the Pacific Ocean to the west. It circulated in the atmosphere around the globe, blocking sunlight and cooling Earth's climate.

What made these super eruptions so explosive is the type of magma that fueled them. And what makes this magma possible is the Yellowstone Hotspot.

The Yellowstone Hotspot begins 1,800 miles below Earth's surface at her core-mantle boundary. From there, a plume of superheated rock rises slowly and begins to pool about 50 miles below the surface. From this 300-mile-wide pool of magma, which raises the land above it almost 2,000 feet, blobs of molten rock hotter than 2,000 degrees Fahrenheit continually rise.

The decay of uranium and other radioactive elements in the Earth's core generates the extreme heat that drives this activity. Earth's inner core blazes at least as hot as the sun's surface—as much as 13,000 degrees Fahrenheit. This great heat radiates through Earth's mantle and towards its cooler surface. Along the way, it fuels the Yellowstone Hotspot, as well as the planet's few dozen other hotspots. It also creates the currents in the mantle that slowly push the continents away from or toward each other.

When basaltic magma surfaces, as it does in the Hawaiian Islands, it generally flows easily and isn't explosive. With the Yellowstone Hotspot, however, the heavier basalt usually remains several miles below the surface and heats the silica-rich crust above it. The resulting rhyolitic magma is much thicker and stickier and over time supercharges with gas.

As pressure builds in a chamber of rhyolitic magma, the land above it bulges upward, and stress fractures eventually form rings around the edges of the bulging land. As more pressure builds, the fractures reach deeper and eventually tap into the chamber, lowering the pressure. The gas in the magma then rapidly expands and explodes to the surface along with the molten rock.

As the vast chamber empties, its roof collapses, and earthquakes continually jolt the area. Over hundreds of thousands of years, smaller flows of rhyolitic and, to a lesser extent, basaltic lava fill in the caldera and obscure its rim. If the hotspot remains active in the area, the magma chamber refills and recharges, and subsequent eruptions blow some of this infill away.

This cycle has been the pattern of the Yellowstone Hotspot, which first exploded 16.5 million years ago on the Oregon-Nevada border, nearly 500 miles from Yellowstone. Since then, more than

100 eruptions have ripped craters into the Earth from that area and into Yellowstone on a roughly northeasterly path. The hotspot itself isn't moving. Rather, the North American continental plate slides an inch or two a year in a southwesterly direction as its eastern edge pushes out from spreading lava in the mid-Atlantic some 5,000 miles away.

When the rhyolitic magma chamber's plumbing weakens in one area after two million years or so, a new center of volcanism emerges to the northeast of the old one. Each center usually explodes several times before tapping out.

When a volcanic center loses its potency, the land above it drops in elevation as much as 2,000 feet, and over hundreds of thousands of years, more nutrient-rich basaltic lava surfaces and smooths over the calderas inside the wide valleys created by them. The eastern side of Idaho's fertile Snake River Plain lies in the trail of the hotspot and foretells what Yellowstone might look like in millions of years when the hotspot lies below eastern Montana. While the broad valley of the eastern Snake River Plain, which lies southwest of the hotspot, is relatively flat and seismically quiet with just an occasional flow of basalt, thousands of minor earthquakes shake the floor of the caldera and the mountains north, east and south of it every year.

Three explosive caldera-forming cycles have struck Yellowstone Country. The first cycle included the three eruptions 2.1 million years ago that left the largest crater, nearly two-thirds of which extends from the central to the southwest area of present-day Yellowstone National Park. The rest of that caldera lies predominantly in Idaho, upslope from the Snake River Plain.

The second caldera-forming eruption, which occurred 1.3 million years ago, blew out a much smaller crater inside of the first one at present-day Island Park, Idaho. The smallest of Yellowstone's three calderas, it measures about a tenth the size of the oldest and ejected about a tenth of the ash and rock. Yet it spewed nearly 300 times as much rock and ash as Mount St. Helens did in 1980.

The youngest crater, and second largest at 30 x 45 miles, sits roughly inside the eastern half of the first one but reaches farther east by about 10 miles. It formed 640,000 years ago from explosions from two vents, one southwest of Purple Mountain near the park's Madison Junction, and one farther to the east near Hayden Valley.

The oldest and youngest caldera-forming explosions destroyed mountains inside Yellowstone and to its southwest, just as the hotspot had done in the Snake River Plain. The southern half of Mount Washburn shared this fate 640,000 years ago; its remaining half abuts the northern edge of the youngest crater. From Mount Washburn to Mount Sheridan, 37 miles to the south, the supervolcano blew away or swallowed the mountains, rocks and soil of the Absaroka Volcanic Supergroup era from millions of years earlier, as well as petrified forests and the fossils of dinosaurs and other animals.

Much of the exposed volcanic rock in Yellowstone today arose from the most recent explosive cycle and includes pyroclastic flows that ran many miles from the volcano's vents. The expanding gas in the thick flows drove them like water across the land, but after the gas discharged, the hellish streams halted. Because of their intense heat, the ash and rock in the flows welded together, forming tuff. Known as the Lava Creek Tuff, this rock ranges in thickness from up to 1,600 feet near the vents to a few feet much farther away. The tuff likely took hundreds or thousands of years to completely cool, and some of it condensed and cracked, forming columns.

Huckleberry Ridge Tuff, the rock type from 2.1 million years ago, wasn't all blown away or buried by the Lava Creek flows and ashfall. It lies exposed in places, including south of Mammoth Hot Springs in the Golden Gate/Bunsen Peak area and at Signal Mountain in Grand Teton National Park.

This older tuff originated from vents located near Ashton, Idaho, Lewis Lake in the southwest of the park, and the Upper Geyser Basin, home to Old Faithful. The flows from these vents ran as far as 60 miles. Some followed the Gardner, Yellowstone, Gallatin and Madison River valleys to the north and northwest. Other flows raced south, reaching present-day Jackson Hole and the Tetons, both of which the hotspot plays a major role in shaping.

For tens of thousands of years after major explosions of the supervolcano, smaller flows of less-explosive rhyolitic lava would erupt and fill in much of the giant craters. Since the last explosion, nearly 30 rhyolitic flows have almost filled the nearly thousand-foot-deep caldera. Some flows traveled as far as 20 miles and hardened into layers a thousand feet thick. Most of these flows remained inside the caldera, but a few ran over the rim. And about a dozen rhyolitic flows erupted beyond the caldera, as well as more than three dozen basaltic ones.

Lava flows also dammed streams, helping to create the basins of Yellowstone, Shoshone, Lewis and Heart Lakes. And the Gibbon and Firehole Rivers, which merge to form the Madison River, largely run between lava flows. Gibbon Falls also cascades over a 1 x 4 mile block of Lava Creek Tuff that dropped from the volcano's rim and into the crater. The Upper and Lower Falls of the Yellowstone River also flow over rhyolitic rock, which erupted during the last explosion.

Lava last flowed from the caldera 70,000 years ago on the Pitchstone Plateau in southwestern Yellowstone. Since then, magma has hardened over and sealed the remaining vents and fractures within the crater as the magma chamber below the surface refills.

Two 1,000-foot-high domes of land over some of the volcano's vents, as well as the caldera floor itself, rise and fall as magma, gas and groundwater circulate through them. These bulges of land, the Sour Creek and Mallard Lake resurgent domes, span roughly 10 miles wide by 7 miles long and lie about 20 miles from each other.

The land rises and falls imperceptibly a few inches a year at most, but the steaming, bubbling and hissing of geysers, hot springs and other thermal features, as well as the hundreds to thousands of earthquakes each year, attest to the giant resting below. And every now and then, the giant rouses, as it did 174,000 years ago when an explosion ripped a 4 x 6 mile hole near present-day Grant Village, which later filled with water, creating Yellowstone Lake's West Thumb.

Above the resting giant, in an area roughly defined by the three calderas, the Yellowstone Plateau rises to about 8,000 feet, pushed up in part by the heat deep below it. Rather than being marked by mountains and valleys, which the explosions blew away or swallowed, the relatively flat landscape features meandering streams and rolling hills—the tops of old lava flows.

CHAPTER 6

Icebound

Yellowstone Country began cooling significantly about 70,000 years ago, around the time of its last lava flow. For centuries, one snowy winter followed another after briefer and cooler summers, and more and more snow remained year-round.

High in the Beartooth, Absaroka and Gallatin Mountains, the snow piled up several hundred feet, compressed and formed thick ice fields. Under pressure from their immense weight and pulled by gravity, glaciers slowly slipped down valleys from these ice fields toward the lower Yellowstone Plateau. They plodded along at rates ranging from less than an inch to as much as 100 yards a day.

Over thousands of years, these creeping rivers of ice merged and thickened on the plateau, forming the Yellowstone Ice Cap. As the ice cap swelled, more and more snow fell on it, and it eventually grew to 4,000 feet thick by 22,000 years ago.

Even outside of glacial periods, deep snows accumulate on the plateau because of the Yellowstone Hotspot. Without the hotspot below, the plateau would be a few thousand feet lower and unlikely to ice over during glaciations. As it is, more than 90 percent of the park is higher than 7,000 feet in elevation.

From the ice ages through today, moist air from the Pacific Ocean follows the track of the hotspot through the eastern Snake River Plain, rises to the plateau and falls heavily as snow during colder months. Much of this snow falls in the southwest area of the park.

The Yellowstone Ice Cap once entombed Mount Washburn, Mount Sheridan and nearly all other peaks in Yellowstone and covered almost 90 percent of the park. Just its western edge remained ice free. This sea of ice and snow extended outward in every direction from the park. Only the highest peaks in the Gallatins, Tetons, Beartooths and Absarokas poked through. A wolverine could have loped north from Jackson Hole to beyond the Beartooth Mountains without ever leaving ice on its 150-mile trek.

Ice ages have occurred about every 100,000 years for the past 2.5 million years when Earth cools and ice spreads from the poles to lower latitudes. Warm periods follow each of these glaciations, with the ice receding toward the poles and mountain tops before spreading again thousands of years later.

Icebound

Long-term patterns of Earth's orbit around the sun play a major role in triggering ice ages. The planet's closeness to the sun in summer versus winter and whether her northern hemisphere tilts toward or away from the sun contribute to the climate. The amount of greenhouse gases in the atmosphere also influences warming and cooling. With lower levels of these heat-trapping gases, more of Earth's warmth escapes into space. Also, as the globe becomes snowier and icier and thereby whiter, more sunlight bounces back into space rather than being absorbed by the darker surfaces of water, rock and soil.

The planet's most recent ice age started about 110,000 years ago. Forty thousand years later, the Yellowstone region's Pinedale glaciation began.

At its peak, the Yellowstone Ice Cap stood out as an enormous island of ice, hundreds of miles south of North America's continental ice sheet. The elevation of the ice cap rivaled many of the peaks from the mountain ranges that had yielded it so much ice, and from its heights, numerous glaciers inched into the lowlands below.

The largest of these outlet glaciers originated from ice flowing off the Yellowstone Plateau. It traveled down the Lamar River drainage and merged with ice flowing from the Beartooth Plateau down the Slough Creek drainage. These glaciers merged west of Lamar Valley and flowed into the Yellowstone River drainage a few miles northeast of Tower Junction. This larger ice flow, the Northern Yellowstone

Outlet Glacier, continued to the northwest through the Yellowstone River corridor inside and then outside of the park. It crept through Yankee Jim Canyon and finally stalled in Paradise Valley to the northwest of Chico Hot Springs, a journey of nearly 90 miles. Ice melted along its slow descent, but the plateaus continually replenished it.

Along its path, other glaciers merged into the Northern Yellowstone Outlet Glacier, like streams flowing into a larger river. Inside the park, glaciers from Hellroaring and Obsidian Creeks joined it. Other outlet glaciers from the Washburn and Gallatin Ranges merged near Bunsen Peak, which sat under 600 feet of ice, and joined the larger outlet glacier downslope from there. Closer to its ending point, more glaciers flowed into it from the Absarokas, north of Yellowstone and east of Paradise Valley, including one from Emigrant Gulch near Chico Hot Springs.

At its maximum, the Northern Yellowstone Outlet Glacier reached a thickness of 1,000 feet near Chico Hot Springs and stretched across the entire Paradise Valley. Other outlet glaciers flowed east into the Absarokas, west down the Madison River drainage and south into Jackson Hole, where the ice scoured out Jackson Lake's basin to 800 feet deep in places.

The glaciers generally followed valleys and carved deeply into their V-shaped walls, breaking off rocks and boulders, and carrying away soil and trees as they slowly grinded along. Meltwater also seeped into cracks and holes in the cliffs above glaciers, where it refroze and broke off rock that fell onto the ice below. Fine-grained to larger rock fragments within the moving ice scoured and polished the walls. Over time, glaciers chiseled the V-shaped river valleys into U-shaped ones, including the Lamar Valley, where that area's outlet glacier easily cut through the crumbly rock from the Absaroka Volcanic Supergroup era.

As the last major sculptors of Yellowstone Country, glaciers also shaped many of its mountain peaks and ridges. In areas where smaller glaciers flowed around three or four sides of mountain tops, they carved horns, including Electric Peak in the Gallatins, the Bear's Tooth in the Beartooths, Pilot Peak in the Absarokas and Grand Teton in the Tetons.

Glaciers carry their debris down to lower and warmer environments. As they move, rocks grind together and create silt, sand and clay particles, as well as even finer glacial flour. The ice gets grittier and grittier the closer it nears its end. Glaciers deposit most of this sediment, along with rocks and boulders, at their edges and ends. Meltwater streaming from the top of and below glaciers pushes much of this material onto outlet plains, such as the ones near Chico Hot Springs, West Yellowstone and Jackson, Wyoming.

On these plains and elsewhere along a glacier's path, a wide-ranging mix of rock settled, from boulders and gravel broken off nearly 3-billion-year-old Beartooth bedrock, to 50-million-year-old supergroup conglomerate, to younger rhyolitic rock and obsidian gravel.

Several times during the older Bull Lake glaciation, which reached its maximum buildup about 140,000 years ago, lava erupted from the Yellowstone Volcano and flowed into ice or ice-cold meltwater. This happened 160,000 years ago when the Nez Perce Creek Flow erupted from a vent in the Mary Mountain area and again about 106,000 years ago between Mammoth Hot Springs and Norris. These silica-rich lava flows cooled quickly amidst billowing clouds of hissing steam, creating obsidian, a black volcanic glass. The younger flow's obsidian extends about half a mile and reaches as much as 200 feet high in places. And for thousands of years, Native Americans and their ancestors quarried this rock at Obsidian Cliff to create tools, including spearheads and arrowheads.

No lava flowed during the more recent Pinedale glaciation, but massive volumes of superheated water, constricted by ice and rock overhead, did explode, leaving craters up to 4,000 feet wide, including Mary Bay and Indian Pond along the northeastern edge of Yellowstone Lake.

In some areas, steam from hydrothermal features melted holes in overlying ice. Meltwater streams on the glaciers deposited sand and gravel into these holes, creating thermal kames, tall hills of sediment that remained after the ice melted, in places such as Yellowstone's Fountain Flats and Mammoth Hot Springs.

Icebound

During an ice age, glaciers retreat and grow, depending on intermittent warming and cooling trends, but between glaciations, the ice melts away altogether except from the highest peaks and from the north and south polar regions.

It took many centuries for the ice to recede from the Yellowstone Plateau and much longer for it to melt from the higher country surrounding it. But by about 12,000 years ago, ice disappeared from all but a few glaciers high in the Tetons and other ranges.

Rivers of meltwater rushed off the Yellowstone Plateau as the climate warmed and its ice broke up. Some of this water worked its way over, under and around melting glaciers in the Lamar and Yellowstone River drainages. The Yellowstone River swelled to as much as half a mile across in Paradise Valley, north of the park.

In several areas, icebergs blocked streams, creating enormous lakes behind them. This included Lamar Lake, which formed when the Slough Creek Glacier crept up the Lamar River and dammed it at Lamar Canyon. Deep waters backed up in the Lamar Valley until they grew high enough to float the ice dam away. A tsunami of towering water then rushed downstream of Lamar Canyon, churning with huge blocks of ice, boulders, rock, mud and sand. The rushing water overwhelmed the Yellowstone River near Tower Junction, careened through the Black Canyon of the Yellowstone River, slammed into the slopes around Gardiner, Montana, and then raced downstream through Yankee Jim Canyon and into Paradise

Valley. Massive floods also rushed down the Madison River toward West Yellowstone and down the Snake River into Jackson Hole after their ice dams broke.

Glaciers dammed and diverted many streams and rivers into their current locations. And they scoured Jackson Lake's basin and Yellowstone Lake's southeast and south arms. Glacial deposits of rock and sediment also dammed the Snake River, filling in Jackson Lake.

In Yellowstone's Tower Junction, Slough Creek and Lamar River areas, retreating glaciers dropped thousands of boulders from the melting ice, some as large as log cabins. In these areas and others, glacial till buried chunks of ice, which later melted, forming kettle lakes and ponds. Some of these exist today, although the warming climate has dried up others.

After the ice retreated from Yellowstone Country, a barren landscape remained. Trees had survived outside of glaciated areas, but elsewhere, glaciers had ripped them and all other plants away along with the soil. At first, a mix of grasses, stunted sagebrush and other plants recolonized the region in the soils formed after the glaciers melted. Willow also returned along stream channels and lakes, and trout and other fish eventually swam upstream into their ancestors' old haunts.

Beavers followed the resurging willow, and other wildlife also moved back into the mountains, valleys and onto the plateau once habitat and food became available.

The returning animals had coexisted with many other species during the ebb and flow of the ice for thousands of years. But most of the animals adaptable enough to make it through several glaciations disappeared not long after the Pinedale ice vanished and humans appeared. This included American cheetahs and lions, short-faced bears, mastodons, mammoths and camels.

By about 13,000 years ago, Engelmann spruce forests began replacing tundra vegetation as the climate continued to warm and remain wet. Other conifers followed the spruce as the climate warmed more. Lodgepoles grew thickly and dominated the forests in the poor soils on the Yellowstone Plateau. In richer andesitic soils below the plateau, mixed spruce, fir and pine forests grew, while in glacial clay areas, including the old lake beds of Hayden and Lamar Valleys, grasslands flourished.

From about 11,500 to 7,000 years ago, the Yellowstone climate warmed to its greatest extent since before the Pinedale glaciation as the sun shone closer to the area during summers. The climate cooled after that and even experienced a minor glacial period, the Little Ice Age, for several hundred years, which ended in the late 1800s.

Since then, the climate has warmed more rapidly than perhaps it ever has. Summers in Yellowstone Country have been growing longer, hotter and drier, and every so often, massive wildfires rage.

The Story of Yellowstone

CHAPTER 7

The Big Burn

Waves of red and orange flames rose hundreds of feet above Yellowstone's forests as columns of thick black smoke billowed five miles into the sky, blocking out the sunlight. Strong winds stoked the fires, launching embers a mile or more ahead of the flames, and spreading them as much as 10 miles a day. Virtually no rain fell for three months that summer in 1988, and the fires failed to subside at night as the humidity remained unusually low. By the time rain and snow dampened the flames in September, 10 fires had blackened more than 2,100 square miles of Yellowstone Country, more than half of which lay in the park.

Lightning ignites fires nearly every year in Yellowstone. The flames usually die out after burning a single tree or small area, thus park managers weren't alarmed when several such fires ignited in

June. But when the right conditions exist, wildfires spread quickly as eventually happened during that parched summer.

Extensive burns in Yellowstone are closely associated with lodgepole pine, the dominant tree in the park's forests, where four of every five trees are lodgepoles. That ratio is even higher on the Yellowstone Plateau, where little else grows in the quick-draining and poor rhyolitic soil. The shallow, horizontally spreading roots of lodgepoles enable them to absorb rain and snowmelt before the water seeps deeper underground and out of reach.

Sun-loving lodgepoles depend on fire to clear the forest floor and to spread their seeds. Indeed, these trees share an extensive history with fire in Yellowstone, going back 11,000 years, about 1,000 years after glaciers had receded from the area. Inside the resin-sealed cones of mature lodgepoles, the seeds of new forests slumber, awaiting the heat of fire to free them. When that heat arrived in 1988, it melted the waxy resin and scattered millions of seeds. The flames also reduced to ash decades-old and even centuries-old accumulations of dead needles, branches and logs on the forest floor—recycling nutrients much quicker and more thoroughly than bacteria, fungi and other decomposers can do in the generally cool, dry Yellowstone climate. And the fires released even more nutrients as they burned live limbs and forest understories.

Forest fires rarely consume all the trees within their perimeters—even fires as intense as in 1988. Damage varies in degree, from fully torched trees to intact ones, creating a mosaic of black, rust, gray and green. Over time, a mix of habitats with greater diversity takes shape, rooted in the varying species and ages of trees and plants. A 25-year-old lodgepole forest hosts more than twice the number of bird, mammal and plant species than one at least twice that age. Living trees, as well as fallen ones and standing deadwood, provide shelter and food for a wide variety of insects, birds and other animals.

In the spring and summer following the fires, moisture from snowmelt and rainfall helped mineral-rich ash seep into the soil. Sunlight warmed the earth, and lodgepole seeds sprouted thickly. In the ensuing years, dense stands of young lodgepoles competed for sunlight, water and nutrients. As larger trees shaded smaller ones and overhanging branches blocked the sun from lower ones, the weak branches died and eventually fell. Over time, more and more trees and branches lost out and died.

The seeds and roots of burned and once-shaded plants also sent up lush new growth during the spring and summer of 1989. Sun-loving Bicknell's geraniums sprouted, their seeds having slumbered for two centuries or more. These pale pinkish-lavender wildflowers grew for two years, dispersed their seeds and then died back, awaiting another fire to be reborn. Also, the pink of fireweed, the yellow of heartleaf arnica and the blueish purple of lupine radiated against the backdrops of blackened trunks in newly opened, sun-drenched areas. Grizzly bears fattened up on the fireweed, clover and other plants in burned areas during the several lush years following the fires.

Dense shoots also emerged from quaking aspen roots after the fires. Some of the aspen groves have root systems dating back more than 10,000 years, making them the oldest living organisms in the area. Aspens typically grow more quickly by cloning themselves from shoots than by sprouting from their tiny, wind-blown seeds. Those seeds must take root in bare patches of soil and then receive consistent moisture with little competition during the next few years. Following the '88 fires, aspens actually did manage to spread to new areas from wind-dispersed seeds, including onto the plateau.

For more than 100 years, aspens have declined in the northern areas of Yellowstone, their stronghold, where they grow at the edges of forests and along floodplains and stream banks. With a drying climate and, until recently, high densities of elk browsing heavily on shoots and young trees, aspens died back to about a third of their former range.

More recently, since the restoration of wolves in Yellowstone, elk numbers have dropped to levels more suitable for the land that supports them, and aspen appear to have benefited from less browsing pressure. As they did with wolves, people also suppressed fire for much of the last century, which allowed conifers to outcompete aspen and close up forests. Today, naturally occurring fires are usually allowed to burn, yet the benefits of fire for aspen could be offset by the rapidly changing climate.

In cooler, wetter areas downslope from the poor soils of the Yellowstone Plateau, shade-tolerant Engelmann spruce and subalpine fir dominate older forests. They grow in the richer soils formed from the andesitic lava of the Absaroka Volcanic Supergroup era. The thin-barked spruce and fir burn easily, but in the absence of fire, they often take root below mature lodgepoles, shade out their seedlings and eventually take over.

Douglas fir also grows in the andesitic soils found off the plateau and has adapted to fire through the eons by growing thick bark that protects it from less intense burns, which occur every 25 to 60 years in the Lamar Valley and other places in the park. The newest of the tree species in Yellowstone Country, Douglas fir seeds blew in about 7,000 years ago. Away from forests, Douglas fir seeds often sprout next to boulders dropped by glaciers, including the thousands scattered about from Tower Junction through the Lamar Valley. The shadows of these nurse rocks retain moisture for the fledgling trees in an otherwise arid landscape.

Whitebark pine grows in steep, rocky ground from an altitude of 7,000 feet up to timberline, where it outcompetes other trees and tolerates fire better than spruce and fir. Whitebarks take many decades to reach maturity and produce pinecones, and cone production varies by area and from year to year. When these cones are available, grizzlies, red squirrels and Clark's nutcrackers feed heartily on the fatty nuts inside. It takes the crushing power of bear jaws, the persistent chewing of squirrel teeth and the stout beaks of jays to crack into the hard cones to reach the nuts.

Some black bears also climb whitebarks for the cones, but grizzlies, who rarely climb trees, let squirrels do the gathering and then raid their caches. Chattering squirrels vigorously defend their stashes of cones, but if they protest too much and too closely, a grizzly will devour them, too. Bears need to eat as much as possible in late summer and autumn before hibernating. They can gain nearly three pounds a day during that period, especially when high-caloric pine nuts are available.

Clark's nutcrackers, like squirrels, also stash seeds for winter and for the next nesting season. At the tops of whitebark trees, nutcrackers strike their chisel-like lower jaw into the seams of the cones to extract the nuts. They stuff as many as 150 pine nuts into their throat pouches and fly off to bury them in caches, sometimes miles apart. Using landmarks, such as logs, rocks and trees, the nutcrackers return to most of their hundreds of caches during the winter and spring to feed themselves and their young. Many of their caches lie in snow-free, windswept areas, but the birds sometimes must sweep away deep snow to reach their stores.

Unlike other jay and crow species, both female and male nutcrackers sit on their eggs, allowing the other to fly off to retrieve the nuts they stored the previous year.

In lean whitebark cone years, nutcrackers in the Yellowstone area seek seeds of the less common limber pines, which are closely related to whitebarks, and they eat just about anything else they can find. Nutcrackers will sometimes fly long distances searching for cones.

New growth of whitebark pine trees depends almost exclusively on nutcrackers forgetting some of their hiding places. Clusters of trees and multi-trunked ones high in Yellowstone Country mark these long-forgotten caches.

The '88 fires burned more than a quarter of the whitebarks in Yellowstone, and since then mountain pine beetles and white pine blister rust, a fungal disease, have sickened and killed an even greater number than the fires affected. Earth's warming climate accelerates these infections, and in the decades to come the trees could vanish from Yellowstone Country and other areas, such as Glacier National Park, where some whitebarks are more than 1,200 years old. Clark's nutcrackers also could disappear since they depend on these trees.

In about an eighth of Yellowstone National Park, far below whitebark country, meadows, sagebrush and grasslands grow in the clay soils left behind by melting glaciers. Had ice not covered the park, the Lamar and Hayden Valleys and other open areas would be thickly forested and home to far fewer elk, bison and pronghorn antelope. About every two decades, fire also burns through these areas and kills small trees that might otherwise establish forests and shrink the grasslands. Although sagebrush, grasses and other plants die back during fires, their roots feed on nutrients from the ash and send up lush new growth afterward.

Just as trees and other plants have adapted to fire, so has wildlife. As the '88 fires raged behind them, elk grazed in smoky meadows and grizzly bears dug for roots in blackened soil. They skirted the approaching flames but remained near good forage, just as they've done for thousands of years. And at least one curious black bear poked at a burning log with its paw, perhaps licking up ants escaping the flames.

Few large animals died in the '88 fires, but the ones that did usually succumbed to choking smoke rather than flames. The carcasses of the elk, bison, moose and deer that died, like ash from the burned forests, yielded nutrients for grizzly bears, coyotes, eagles, ravens and others. Carnivores also devoured the squirrels, chipmunks, mice, voles and other small animals fleeing the flames. And burned areas provided less cover for prey and easier hunting for coyotes, foxes, weasels, hawks and more.

Of all the animals, moose experienced the greatest long-term effects from the '88 fires. While just a few of them died that summer, much of their winter forage of willow and old-growth subalpine fir burned. This led to as much as 75 percent fewer moose living in the park today.

Moose migrated into the Yellowstone area relatively recently, however, compared to other species. They moved into the south of the park by the 1880s and into the north a few decades later. They also seemed to have benefited from the lack of fire until 1988.

The severe drought that summer and the severe winter that followed took their toll on other hoofed animals besides moose, at least in the short term. With meager grazing and browsing opportunities that summer and fall, and with deep snow and frigid temperatures that winter, many ungulates starved to death. Nearly a quarter of the park's mule deer and pronghorn antelope died, about 40 percent of northern Yellowstone's nearly 19,000 elk perished, and close to 20 percent of the park's bison died. In many cases, hunters shot the hungry elk when they left the park looking for forage, and wildlife managers and hunters shot many of the bison, as well.

However, tough winters for hoofed animals lead to fat springs for hungry grizzlies emerging from their dens, as well as for coyotes, foxes, ravens, eagles and other scavengers. Also, the 1988–89 winter followed two mild ones with relatively few deaths, and the herds of most animals increased in the years afterward, although moose numbers did not.

Like moose, boreal owls lost much of the old-growth forests on which they rely after the fires. Similarly, some osprey young, still in their nests, burned in the flames, but many bird species benefited from the fires. Cavity-nesting birds found plenty of homes to choose from in burned trees. These include Barrow's goldeneyes, kestrels, flickers, tree swallows and mountain bluebirds. Dead trees also teemed with insects and larvae for hungry flycatchers and woodpeckers. And open ground below blackened trunks provided more opportunities for ground-nesting birds and for flickers, robins and other thrushes hunting insects and worms.

Seed-eating birds, including Clark's nutcrackers, finches, crossbills and pine siskins, gobbled up many of the lodgepole seeds dispersed by the fires, but not enough to prevent the growth of new forests. These forests have regreened the blackened landscape, while countless weathered trunks of burned trees stand as a reminder of that year's widespread fires.

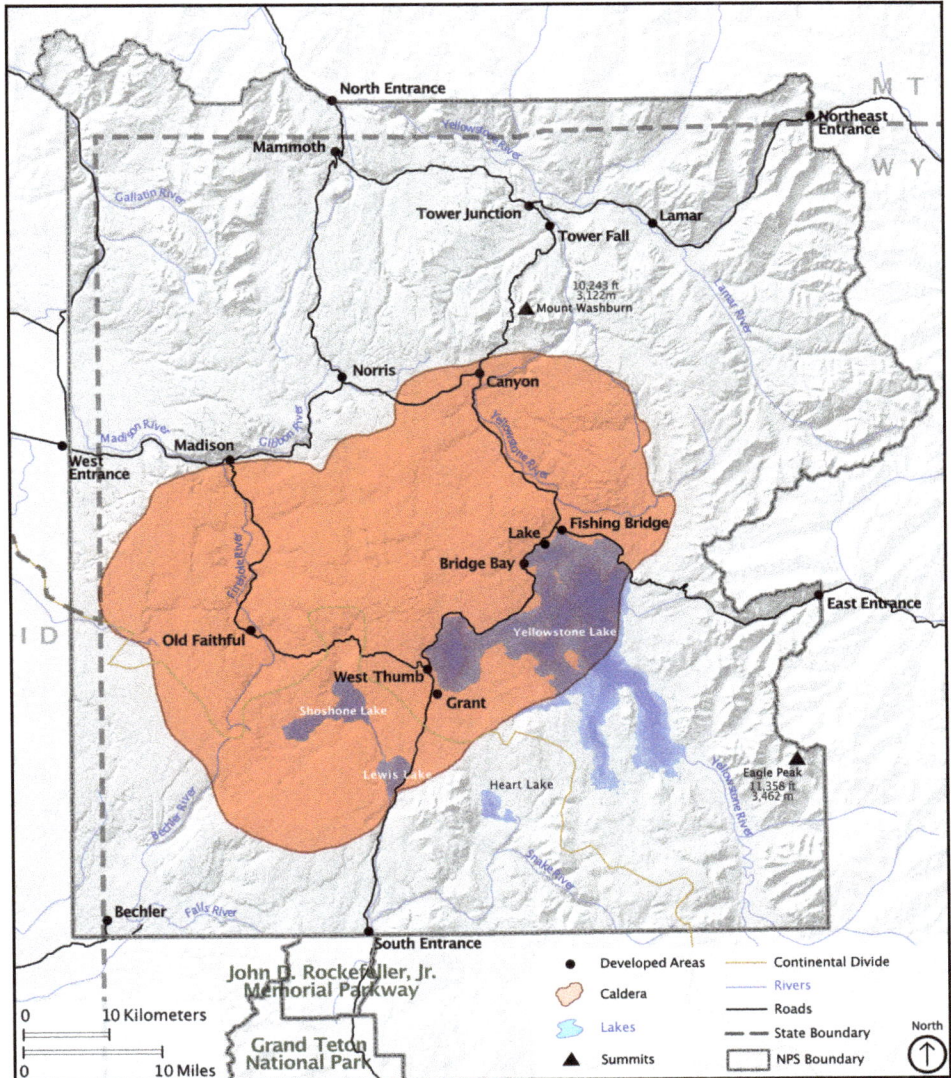

Simplified map of Yellowstone National Park. Adapted from National Park Service (NPS)

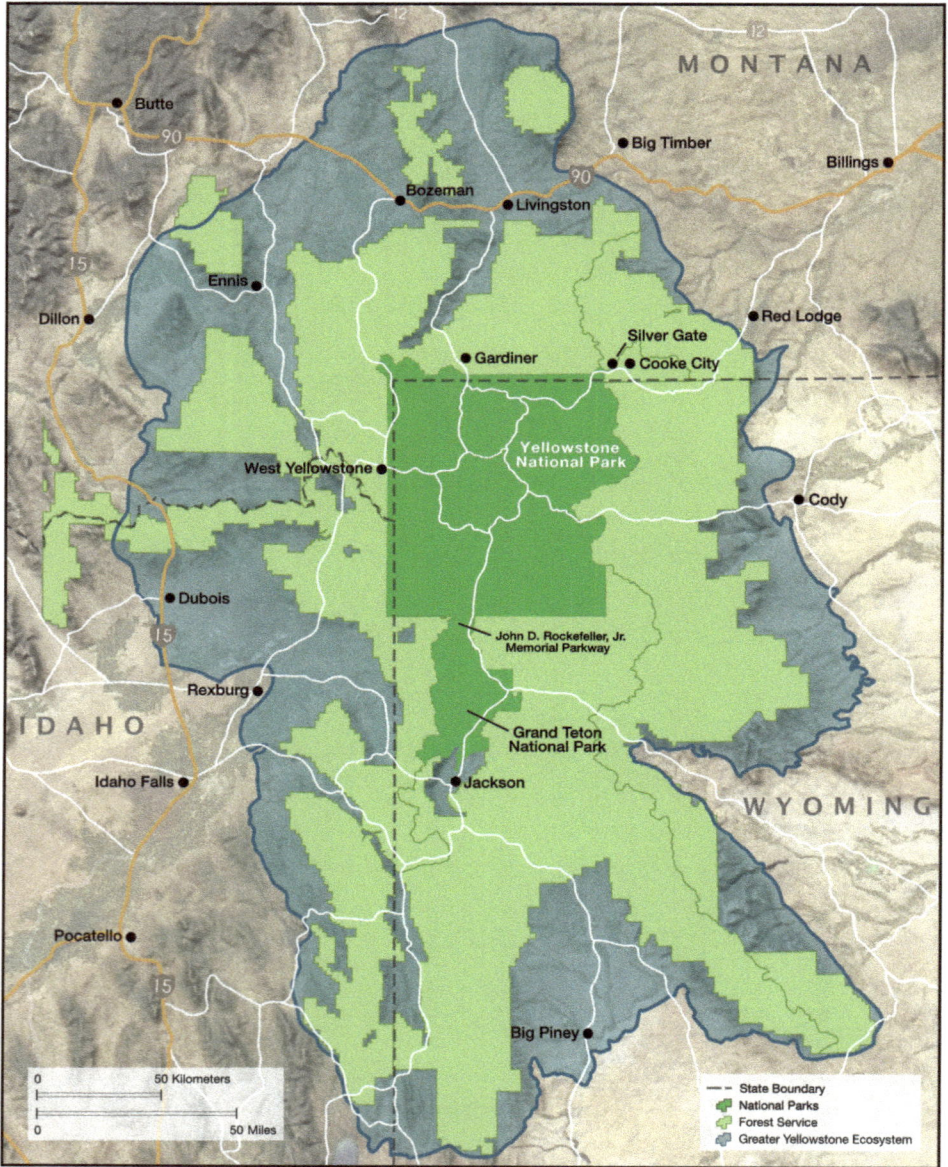

The Greater Yellowstone Ecosystem (GYE) or "Yellowstone Country," The borders for the GYE vary depending on the entity defining it. This map shows it at its maximum—some 10 times larger than Yellowstone National Park itself. Adapted from Lisa Landenburger, US Geological Survey (USGS) and NPS

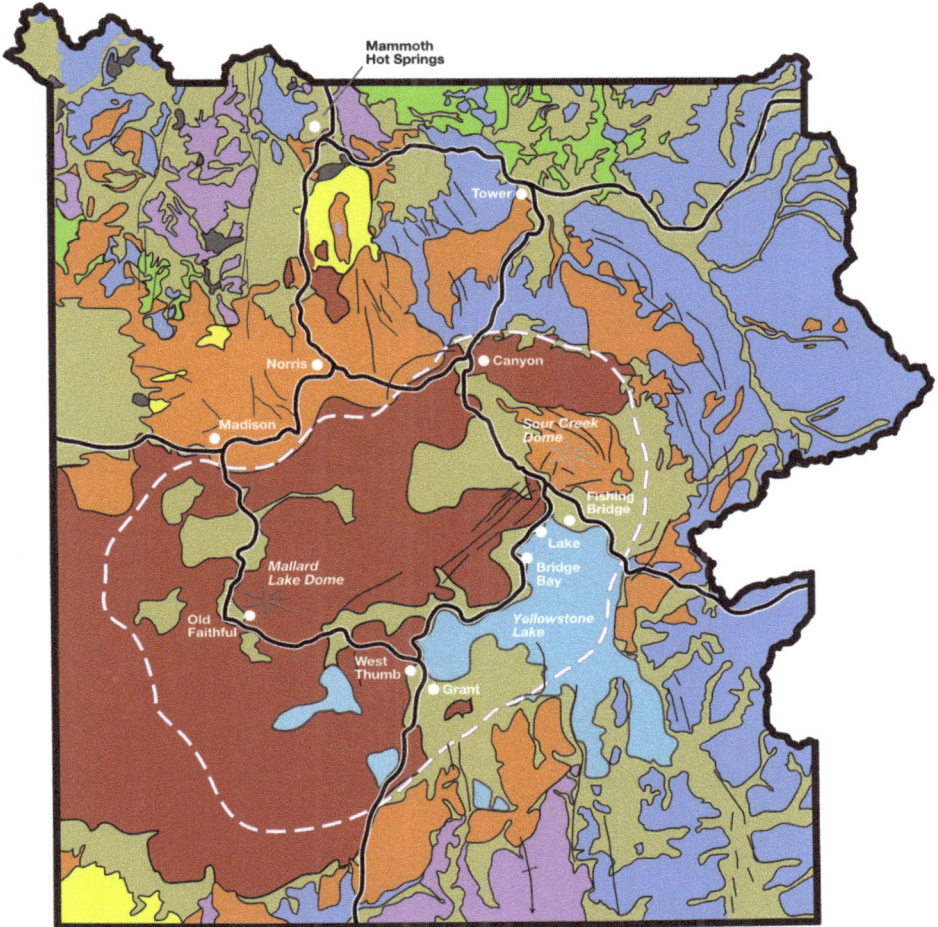

Simplified geologic map of Yellowstone National Park. Adapted from USGS and Marli Bryant Miller, University of Oregon

Map labels:

Mammoth Hot Springs

Tower

Norris

Canyon

Sour Creek Dome

Madison

Fishing Bridge

Lake

Mallard Lake Dome

Bridge Bay

Old Faithful

Yellowstone Lake

West Thumb

Grant

Legend:

Stream and glacial deposits, and thermal areas
Within past 17,000 years

Rhyolite lava flows (after major volcanic eruptions)
2.1 million to 72,000 years old

Basalt lava flows
Within past 2.1 million years

Welded tuff from supervolcanic eruptions
~2.1 million years and 640,000 years old

Absaroka Volcanic Supergroup rock
~50 million years old

Absaroka Volcanic Supergroup igneous intrusions
~50 million years old

Sedimentary rocks
~83-542 millions years old

Gneiss, granite and schist bedrock
~2.7 billion years old

Yellowstone caldera

Faults

Fold

Resurgent dome

Paved road

Eon	Era	Period	Epoch	MYA		Life Forms	North American Events
Phanerozoic	Cenozoic (CZ)	Quaternary (Q)	Holocene (H)	0.01	Age of Mammals	Extinction of large mammals and birds Modern humans	Ice age glaciations; glacial outburst floods
			Pleistocene (PE)				
				2.6			Cascade volcanoes (W) Linking of North and South America (Isthmus of Panama)
		Neogene (N)	Pliocene (PL)			Spread of grassy ecosystems	Columbia River Basalt eruptions (NW)
			Miocene (MI)	5.3			Basin and Range extension (W)
			Oligocene (OL)	23.0			
		Paleogene (PG)	Eocene (E)	33.9			Laramide Orogeny ends (W)
			Paleocene (EP)	56.0		Early primates	
				66.0		Mass extinction	
	Mesozoic (MZ)	Cretaceous (K)			Age of Reptiles	Placental mammals	Laramide Orogeny (W) Western Interior Seaway (W)
				145.0		Early flowering plants	Sevier Orogeny (W)
		Jurassic (J)				Dinosaurs diverse and abundant	Nevadan Orogeny (W) Elko Orogeny (W)
				201.3		Mass extinction First dinosaurs; first mammals Flying reptiles	Breakup of Pangaea begins
		Triassic (TR)					
				251.9		Mass extinction	Sonoma Orogeny (W)
	Paleozoic (PZ)	Permian (P)			Age of Amphibians		Supercontinent Pangaea intact
				298.9		Coal-forming swamps Sharks abundant First reptiles	Ouachita Orogeny (S) Alleghany (Appalachian) Orogeny (E)
		Pennsylvanian (PN)					Ancestral Rocky Mountains (W)
				323.2			
		Mississippian (M)			Fishes		
				358.9		Mass extinction First amphibians First forests (evergreens)	Antler Orogeny (W) Acadian Orogeny (E-NE)
		Devonian (D)					
				419.2			
		Silurian (S)				First land plants Mass extinction	
				443.8	Marine Invertebrates	Primitive fish	Taconic Orogeny (E-NE)
		Ordovician (O)				Trilobite maximum	
				485.4		Rise of corals	Extensive oceans cover most of proto-North America (Laurentia)
		Cambrian (C)				Early shelled organisms	
				541.0		Complex multicelled organisms	Supercontinent rifted apart
	Proterozoic	Precambrian (PC, W, X, Y, Z)					Formation of early supercontinent Grenville Orogeny (E)
				2500		Simple multicelled organisms	First iron deposits Abundant carbonate rocks
	Archean					Early bacteria and algae (stromatolites)	Oldest known Earth rocks
				4000			
	Hadean					Origin of life	Formation of Earth's crust
				4600		Formation of the Earth	

A timeline of Earth's history. **NPS**

An artist's rendition of early Earth's surface with a molten moon in the background. The image also depicts streaking meteorites that frequently bombarded Earth and the moon early on and occasionally still do. In Yellowstone today, magma remains relatively close to the surface. Tim Bertelink, *Artist's Impression of the Hadean Eon*, 2016

A 20-foot-wide "skylight" in hardened basalt with lava flowing beneath it. Early in Earth's history, she was covered by a basalt crust that was routinely punctured by asteroids and meteorites. Photo taken at Hawaii's Kilauea volcano in 2016. USGS

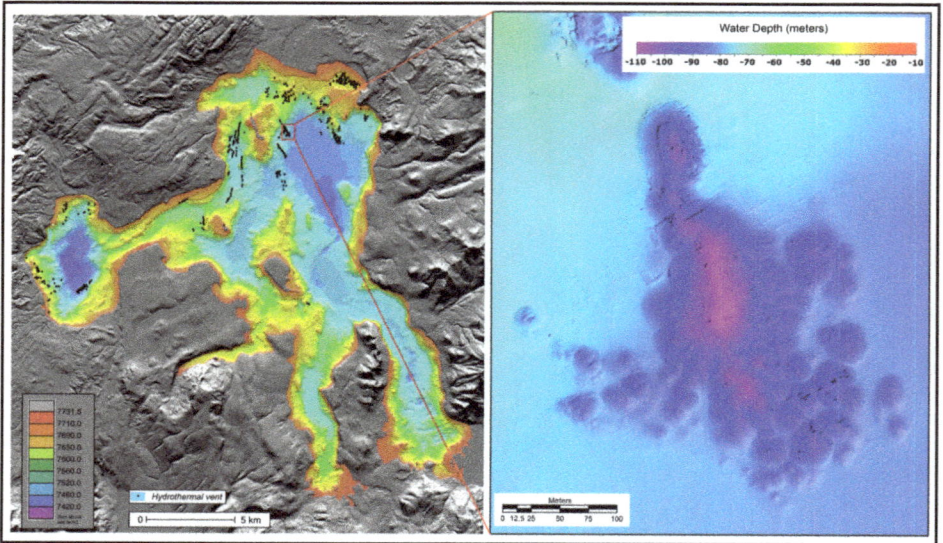

High-resolution depth map of Yellowstone Lake (left) showing hydrothermal vents (black dots) on the lake floor. These vents are similar to the ones from which life on Earth is thought to have originated. Detail at right is of an area known as "Deep Hole." USGS

Grand Prismatic Spring, Yellowstone's largest hot spring. Some heat-loving microbes in the park's thermal areas are similar to the ones that first arose on Earth more than four billion years ago. Jim Peaco, NPS

Map showing the supercontinent Rodinia starting to break up 750 million years ago. Yellowstone is in red. © 2013 Ron Blakey, Colorado Plateau Geosystems Inc.

Map showing Yellowstone south of the equator and under a shallow tropical sea 500 million years ago. © 2013 Ron Blakey, Colorado Plateau Geosystems Inc.

Western tiger salamanders live in wetter areas of Yellowstone. As amphibians, they're an evolutionary link between fish and reptiles, the latter of which came to dominate Earth for nearly 200 million years. **Jacob W. Frank, NPS**

Mammoth Hot Spring's terraces are created from limestone particles bubbling up to the surface. That limestone formed during millions of years from the accumulation of shelled creatures on the shallow, tropical seafloor that later rose up. **Diane Renkin, NPS**

Map showing Yellowstone 315 million years ago north of the equator and still under the sea. © 2013 Ron Blakey, Colorado Plateau Geosystems Inc.

Map showing Yellowstone under a shallow sea at the beginning of the dinosaurs' reign 245 million years ago. Marine reptiles, sharks, crocodiles and rays lived in that sea while dinosaurs roamed the nearby coastal plains to the east. © 2013 Ron Blakey, Colorado Plateau Geosystems Inc.

Ichthyosaurs, marine reptiles that lived throughout Earth's oceans during the dinosaur era, resembled dolphins and filled a similar ecological niche. This image depicts the European *Stenopterygius* from 150 million years ago during the Jurassic period. Adults and juveniles are feeding on small cephalopods. © Douglas Henderson, *Jurassic Ichthyosaur*, 1992

About 180 million years ago, the sea had receded from Yellowstone Country, and the supercontinent Pangaea was breaking up as Africa and South America moved away from North America. At least two species of long-necked dinosaurs, the largest animals ever to walk the Earth, lived in present-day Yellowstone Country during this time.

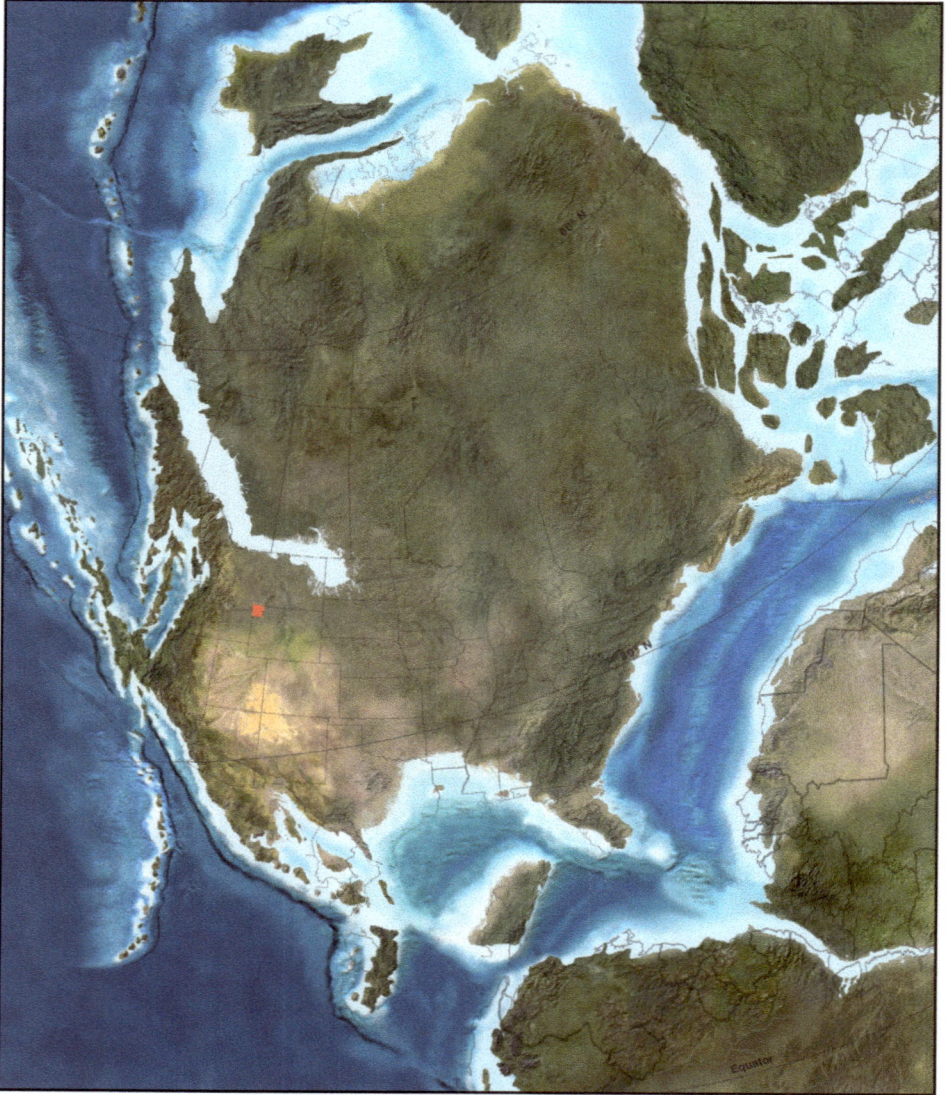

Near the end of the Jurassic 150 million years ago, Africa and South America continued moving away from North America. At this time, *Stegosaurus*, *Allosaurus* and long-necked dinosaurs lived in the Yellowstone area. © 2013 Ron Blakey, Colorado Plateau Geosystems Inc.

Two Allosaurs pursue a *Diplodocus* in a Colorado flood plain during the late Jurassic, some 150 million years ago. A smaller dinosaur (*Ornitholestes*) looks on while a few pterosaurs (*Rhamphorhynchus*) fly by. © Douglas Henderson, *Diplodocus and Two Allosaurs*, 2002

Stegosaurs forage in a late Jurassic landscape of conifers, ginkgo and cycads. © Douglas Henderson, *Stegosaurs Ambiance*, 1986

Map showing the Western Interior Seaway 85 million years ago. The sea stretched from the Arctic to the Gulf of Mexico and ebbed and flowed from Yellowstone during the Cretaceous, the last period of the dinosaurs' reign. The mountains forming to the west of Yellowstone eventually pushed away the sea by 83 million years ago. © 2013 Ron Blakey, Colorado Plateau Geosystems Inc.

Maiasaurs at the edge of the Western Interior Seaway (the "Niobrara"). © Douglas Henderson, *Maia Reaches the Niobrara*, 1983

A group of Maiasaurs on the move. Other dinosaurs, including *Triceratops* and *Edmontosaurus*, are also thought to have formed large herds, not unlike elk and bison today. © Douglas Henderson, *Maiasaura Herd*, 1983

Map showing the Western Interior Seaway 77 million years ago. The sea by then was several hundred miles east of the lush Yellowstone area, where *T. rex* and *Triceratops* lived. © 2013 Ron Blakey, Colorado Plateau Geosystems Inc.

A tyrannosaur in Montana from the late Cretaceous. © Douglas Henderson, *Tyrannosaur Eden*, 1998

Two *Triceratops* spar beyond a group of smaller dinosaurs (*Struthiomimus*) in a Montana landscape some 67 million years ago. © Douglas Henderson, *Struthiomimus and Triceratops*, 2000

A small mammal foraging in a late Cretaceous forest in Montana. © Douglas Henderson, *Mammal and Leaves*, 1990

The asteroid that will mark the end of the dinosaurs' reign speeds toward Earth.
© Douglas Henderson, *Final Approach*, 1999

A mammal surveys a decimated forest as clearing skies allow sunlight to return following the asteroid's impact. © Douglas Henderson, *Surviving Mammal*, 1999

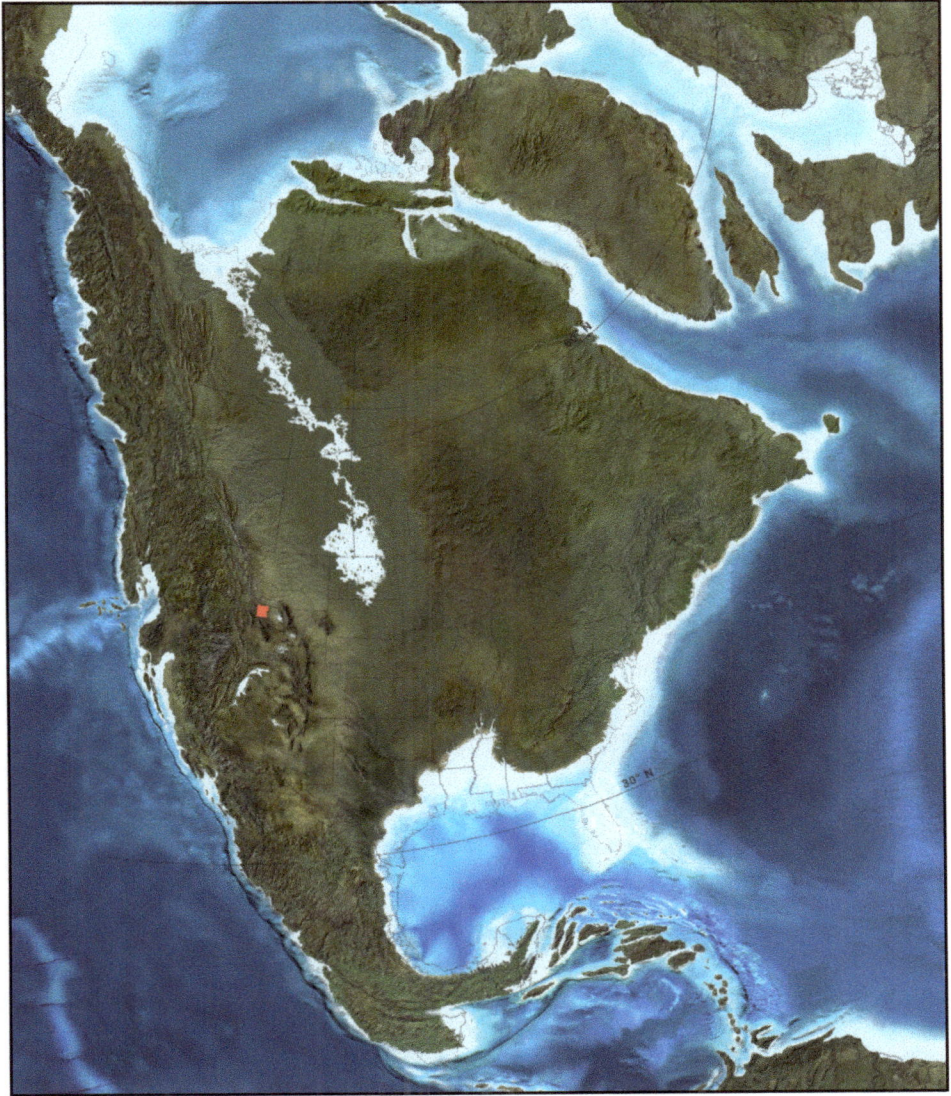

As mammals began their rise about 65 million years ago, the great sea between the Arctic and the Gulf of Mexico had largely disappeared, and mountains continued to form near Yellowstone. © 2013 Ron Blakey, Colorado Plateau Geosystems Inc.

Two saber-toothed cats (*Smilodon*) approaching a bison. The species portrayed in this drawing has been extinct for thousands of years. © Douglas Henderson, *Buffalo and Cats*, 2011

Advancing ice will eventually scrape away the forests and tundra of Yellowstone's Blacktail Plateau and force Mastodons and other animals to lower elevations. © Douglas Henderson, *Advancing Ice on the Blacktail*, 2014

OPPOSITE PAGE: A small group of mammoths ford a meltwater stream along the huge front of the receding Northern Yellowstone Outlet Glacier near the current site of Gardiner, Montana. © Douglas Henderson, *Glacier Front and Mammoth*, 2014

Unlike 75 percent of other mammals living in North America 13,000 years ago, bighorn sheep still exist and live in the Yellowstone area. Diane Renkin, NPS

Along with the ancestors of Native Americans, elk moved into North America from Beringia as the last ice age melted away. Neal Herbert, NPS

The Beartooth Mountains arose to the northeast of Yellowstone some 60 million years ago. The "Bear's Tooth," which glaciers sculpted, is in the lower center of the photo. Jacob W. Frank, NPS

The Hoodoo Basin in northeastern Yellowstone is sculpted from the crumbly rock of the Absaroka Volcanic Supergroup era. Jacob W. Frank, NPS

An astronaut aboard the International Space Station photographed the Sarychev Volcano in the early stages of eruption in 2009. A pyroclastic flow appears to be hugging the ground as it descends from the island volcano's summit. Similar volcanic eruptions some 50 million years ago took place repeatedly throughout Yellowstone Country during the Absaroka Volcanic Supergroup era. **NASA's Earth Observatory**

A close-up image of ash erupting more than 12 miles into the atmosphere from Mount St. Helens, a stratovolcano like the ones that dominated the Yellowstone area during the Absaroka Volcanic Supergroup era. **Donald A. Swanson, USGS**

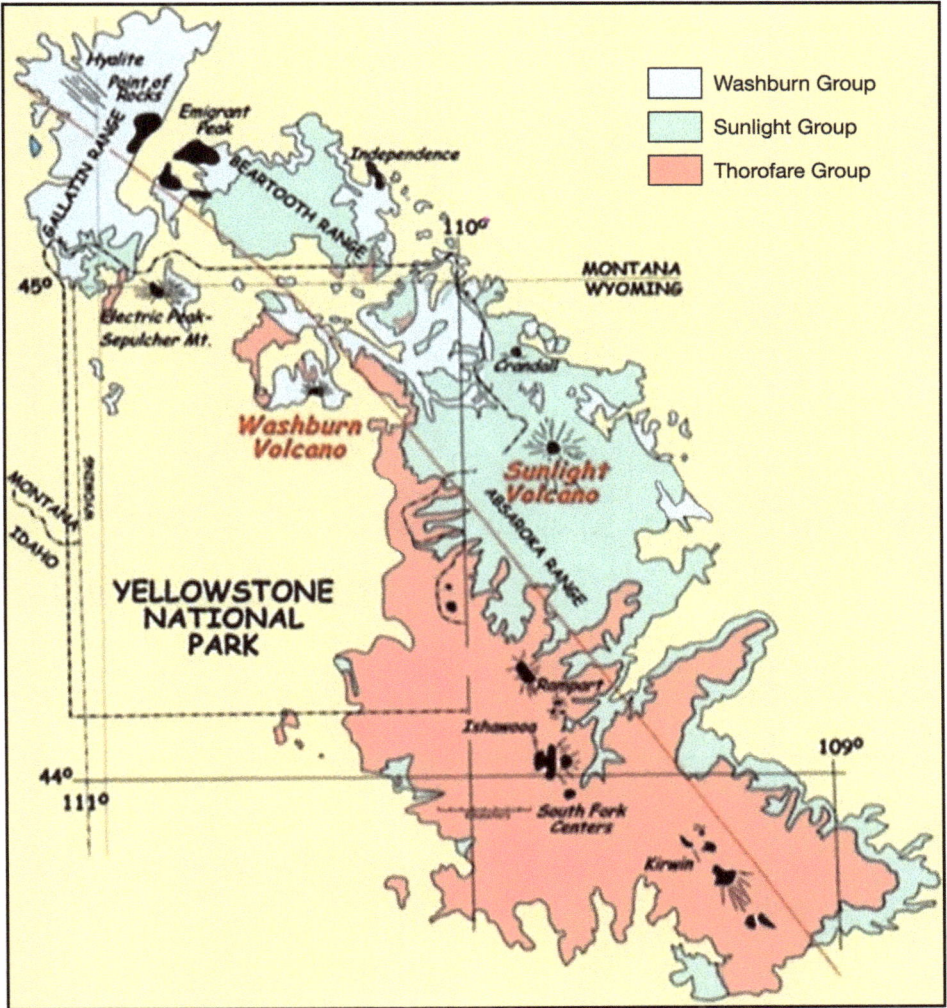

The three groups of the Absaroka Volcanic Supergroup (AVS) era. The Washburn Group is the oldest, the Sunlight Group is second oldest, and the Thorofare Group is the youngest. Younger rock from the AVS era lies on top of older rock in places. Areas within Yellowstone in light yellow represent parts of the park in which AVS rock was eroded, overlain by younger rock or, as in the case of the southwest of the park, blown away by supervolcanic eruptions millions of years later. Modified from USGS, Smedes and Prostka (1972) and *Geology of Wyoming*

The Absaroka Range runs along the eastern side of Yellowstone. Streams, glaciers and other agents of erosion have carved deeply into its soft, crumbly rock for millions of years. Neal Herbert, NPS

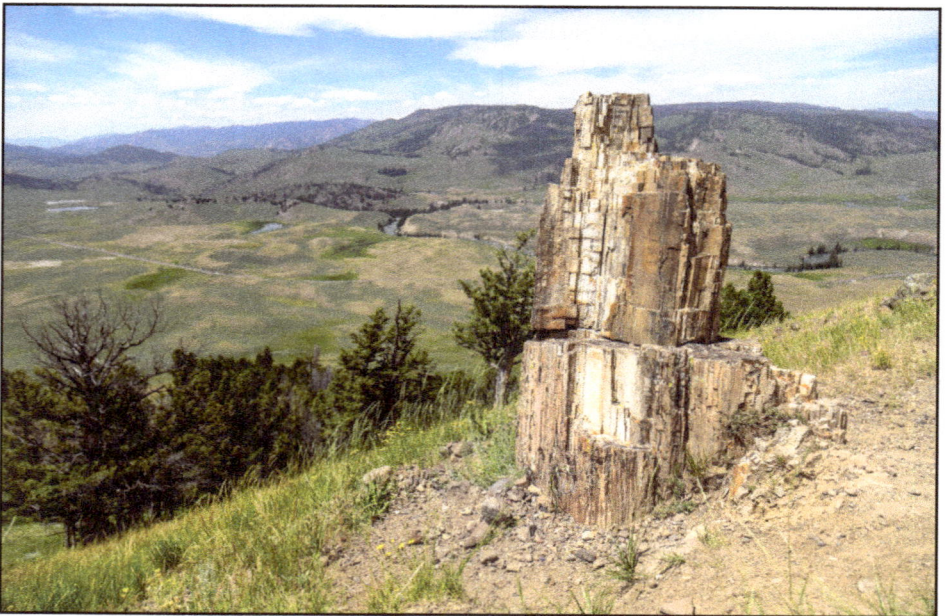

A petrified redwood tree on Yellowstone's Specimen Ridge. Eruptions during the Absaroka Volcanic Supergroup era buried forests in thick flows of ash, mud and grit, in which many trees later fossilized. Jacob W. Frank, NPS

Map of present-day North America showing the Yellowstone area after its mountains fully formed. © 2013 Ron Blakey, Colorado Plateau Geosystems Inc.

LOWER HALF OF NEXT PAGE: Map showing the 500-mile track of the Yellowstone Hotspot. It first surfaced along the Nevada-Oregon border 16.5 million years ago and now lies below Yellowstone. The orange outlines show the chief centers of volcanic activity, including the three massive eruptions in the Yellowstone area. ("Ma" stands for millions of years ago.) USGS

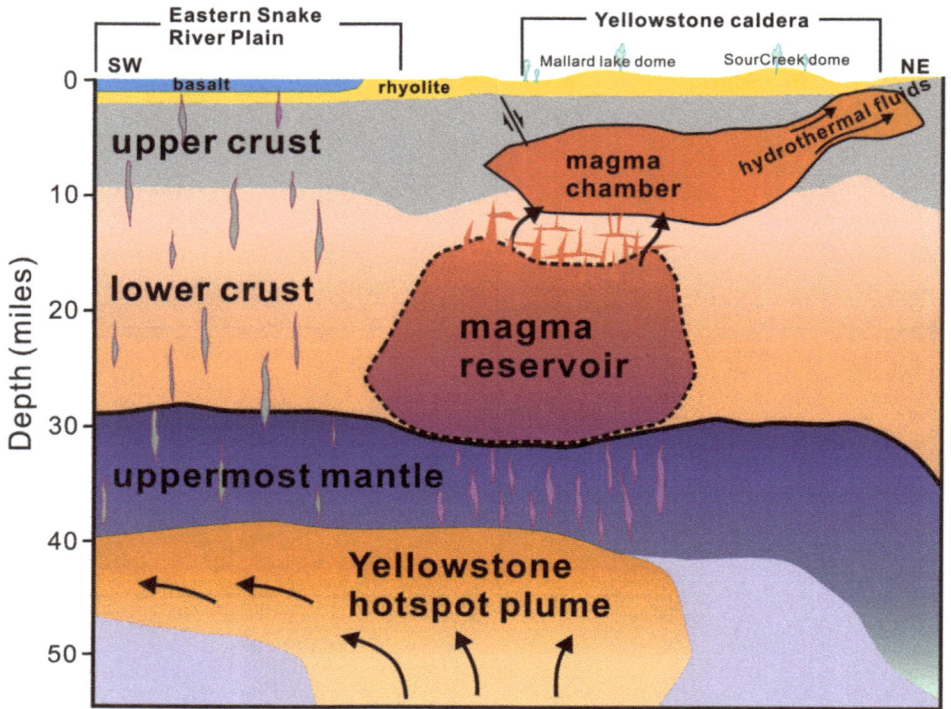

Illustration showing the Yellowstone Hotspot at the bottom (in orange) and the two chambers of partially molten rock (in red). Huang et al., University of Utah

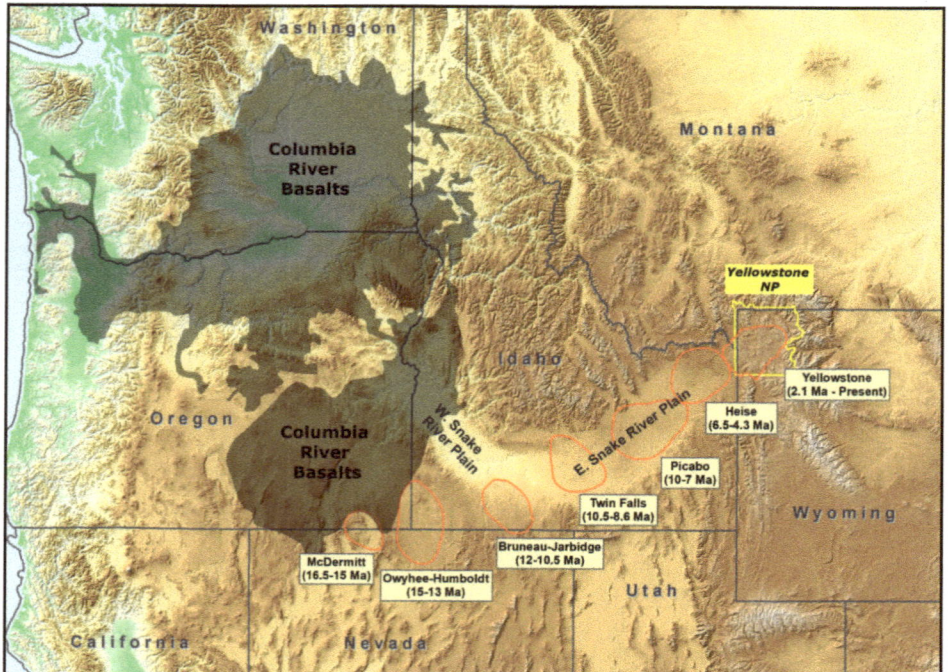

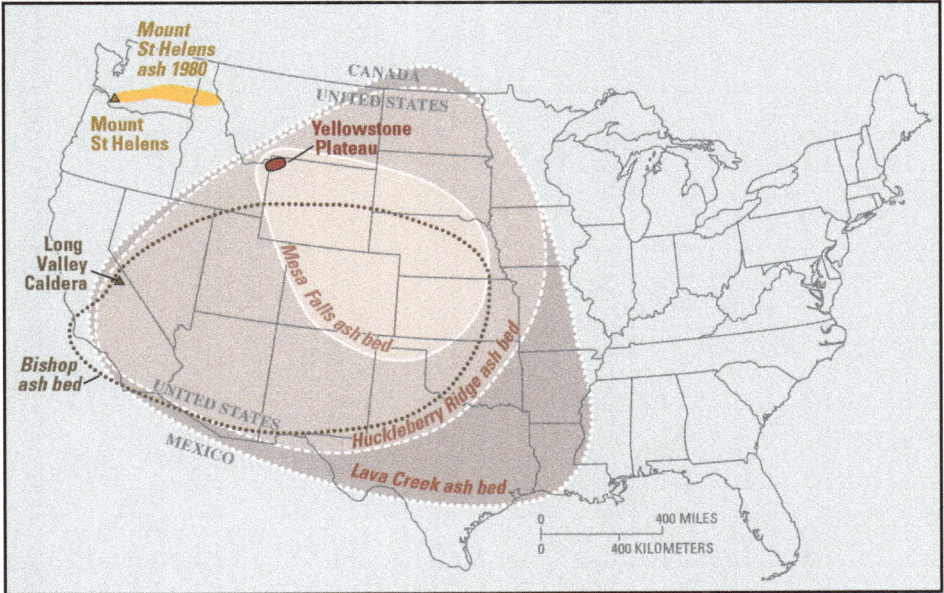

Map showing the known ash-fall boundaries for several US volcanos, including the three Yellowstone super eruptions (the Huckleberry Ridge, Mesa Falls and Lava Creek ash beds). USGS

Vehicles provide scale for the Golden Gate area a few miles south of Mammoth Hot Springs. Just above the vehicles on the left, two distinct flows of Huckleberry Ridge Tuff can be seen. These lava flows raced down the Gardner River drainage 2.1 million years ago. Jacob W. Frank, NPS

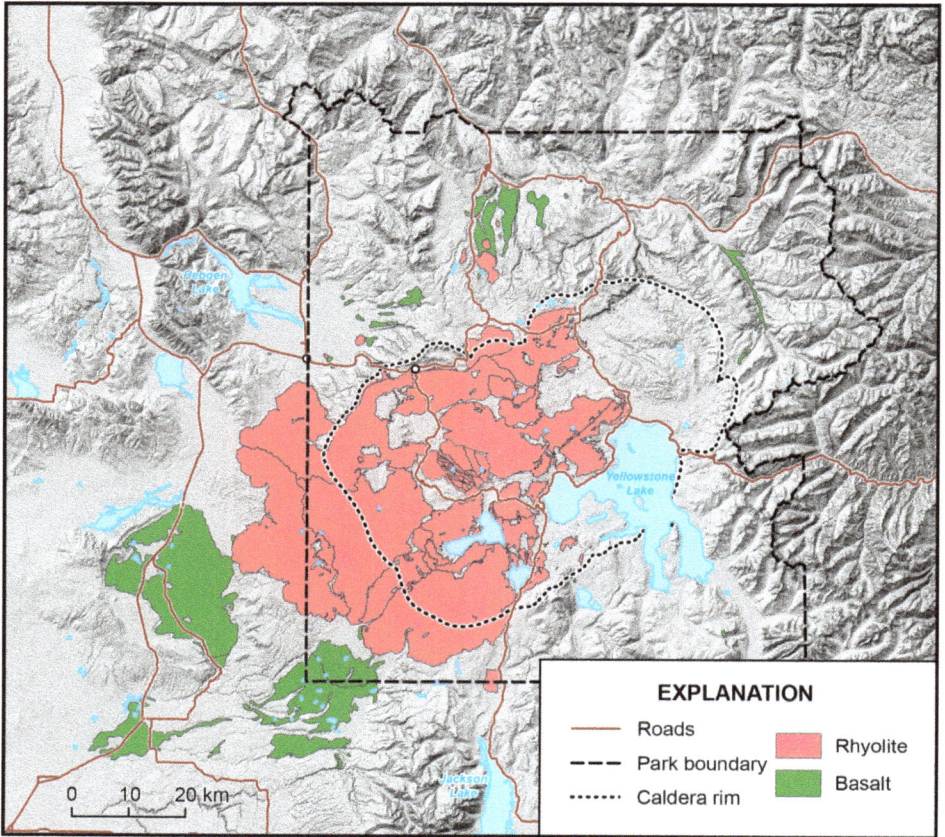

Map showing the distribution of rhyolitic lava that erupted after the formation of the youngest Yellowstone caldera and the basaltic lava that erupted outside the caldera.
Adapted from USGS

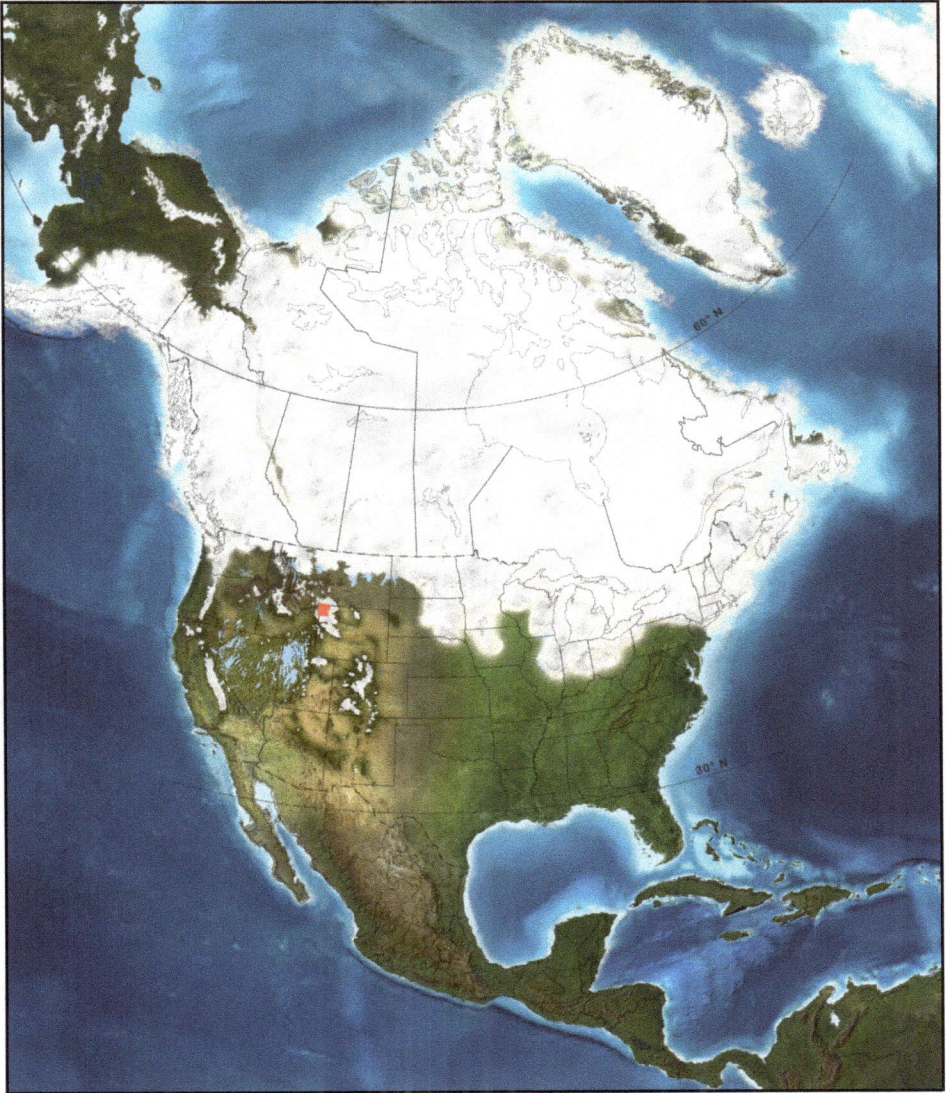

About 22,000 years ago, much of Yellowstone Country was one of several islands of ice beyond the mass of glaciers covering Canada and the northern United States. The large continental ice sheet blocked passage from the mostly ice-free land bridge between Eurasia and North America. Both wildlife and the ancestors of Native Americans lived on that land (Beringia) and would eventually move southward. © 2013 Ron Blakey, Colorado Plateau Geosystems Inc.

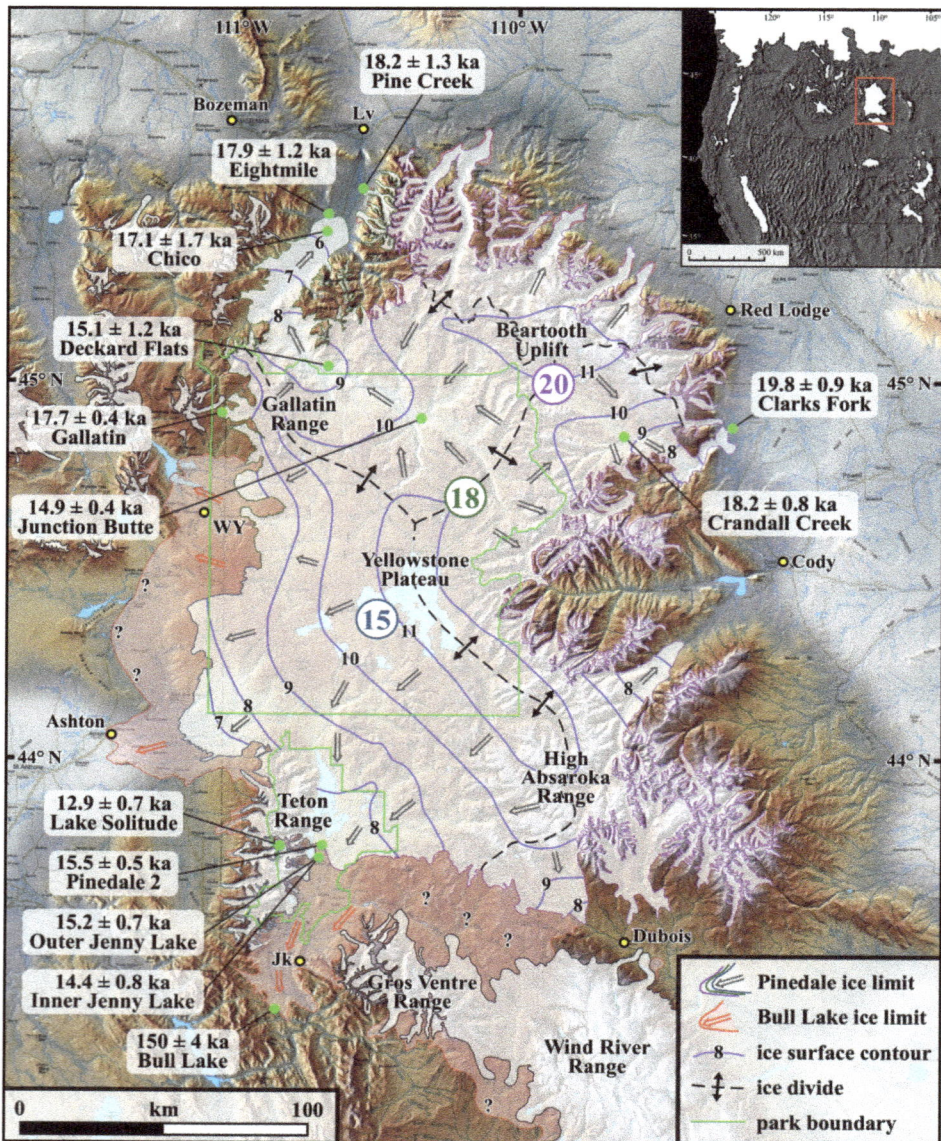

Map showing ice cover in the Yellowstone region from the last two ice ages. Light shaded areas inside black lines indicate land covered during the Pinedale glaciation; red outlined areas show the extent of land covered by the older Bull Lake glaciation. Purple lines are contours in thousands of feet for the height of the ice surface during the Pinedale glaciation. The circled numbers depict the southwest migration of the center of the glacial system through time (in thousands of years–"ka" on map). Joe Licciardi and Ken Pierce, USGS

Glaciers carved The Grand (center) and these other peaks in the Teton Range. **Author photograph**

This photograph from near Beartooth Pass shows the U-shaped sweep that glaciers leave behind while moving downslope. **Jacob W. Frank, NPS**

By 13,000 years ago, the ice was receding in North America, and a wide route from the Arctic down into the Yellowstone area and beyond allowed wildlife and the ancestors of Native Americans to travel southward from the north. The first people to come to Yellowstone, however, might have followed rivers from the Pacific Coast to reach the area. © 2013 Ron Blakey, Colorado Plateau Geosystems Inc.

Morning on Bunsen Peak with valley glaciers and hot springs, viewed from slopes of Mount Everts. © Douglas Henderson, *Morning on Bunsen Peak*, 2015

After glaciers scoured the landscape and waterways in the Yellowstone region, pioneering animals, like cutthroat trout, returned from lower areas that had remained ice-free. Neal Herbert, NPS

A wind-driven fire burns sagebrush on Yellowstone's Blacktail Plateau in 1988, two nights before snow and rain dampened the flames for good that year. Jim Peaco, NPS

A view of lodgepole pines from the forest floor. Lodgepoles generally grow straight in Yellowstone, and their lower sun-starved branches drop off over time. Native Americans used younger trunks as tepee ("lodge") poles. Jacob W. Frank, NPS

Fire mosaic from the 2003 Bacon Rind Fire, which burned in the northwestern area of Yellowstone and adjoining public lands. **Author photograph**

Young lodgepoles growing amid weathered tree trunks that burned in 1988. **Jim Peaco, NPS**

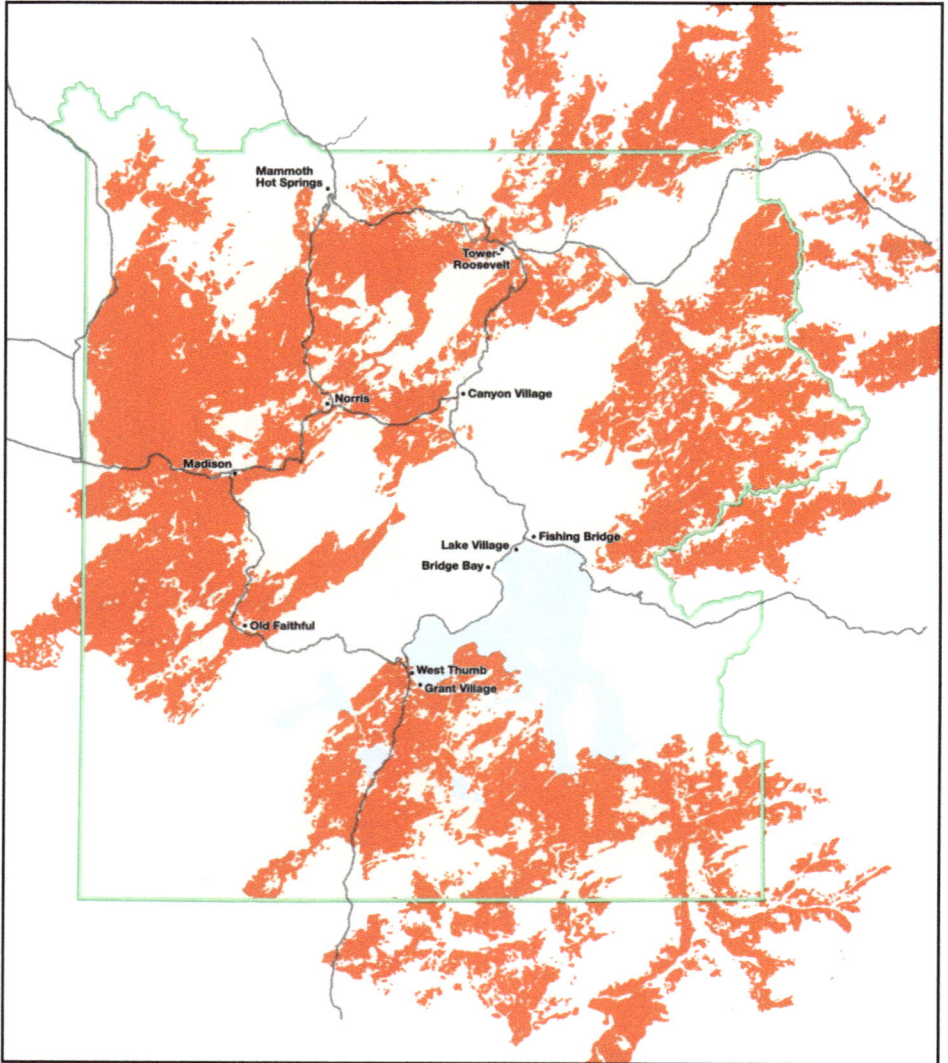

Map showing the fires that burned more than a third of Yellowstone in 1988. **Adapted from NPS**

A mixed aspen-conifer forest in Custer Gallatin National Forest north of Yellowstone. Aspen declined in Yellowstone in the twentieth century because of heavy browsing by elk and fire suppression by park managers, which allowed conifers to outcompete sun-loving aspens. Diane Renkin, NPS

Two bison graze in a smoky meadow in 1988 as a ground fire burns nearby. Jeff Henry, NPS

Grizzly bears are omnivores and benefitted from various food sources after the '88 fires and the following severe winter. Foods included the carcasses of elk, bison and deer and the lush growth of fireweed, clover and other plants. Jim Peaco, NPS

Yellowstone's moose population declined significantly after 1988 when that year's fires burned large swaths of mature subalpine fir, their preferred winter forage. Jim Peaco, NPS

Clark's nutcrackers are closely linked to whitebark pines, which have dramatically declined in recent decades in Yellowstone because of fire, disease and beetle infestations. Jim Peaco, NPS

Mountain bluebirds are some of the many birds that benefit from fires, which create new and varied habitat. Neal Herbert, NPS

Map showing Beringia, from which countless species, including dinosaurs, mammoths and people, moved from Eurasia to North America until about 10,000 years ago. By that time, water from melting glaciers had swelled the oceans and flooded Beringia. **NPS**

Obsidian Cliff, an important source of stone for tool making for Native Americans and their ancestors for more than 11,000 years. Daniel Mayer

A ceremonial obsidian spearhead from a Hopewell Mounds burial site in Ohio. The spearhead dates back 2,000 years and is part of the 300 pounds of Yellowstone obsidian found at the site. NPS

Wickiups at Wickiup Creek in the Gallatin Mountains in 1973. Harlan Kredit, NPS

A photograph of a Sheep Eater (Tukudika) family taken by William Henry Jackson in 1871. **NPS**

An 1871 photograph of Washakie and his people. Washakie was the chief of the Eastern Shoshone in the last half of the nineteenth century. **NPS**

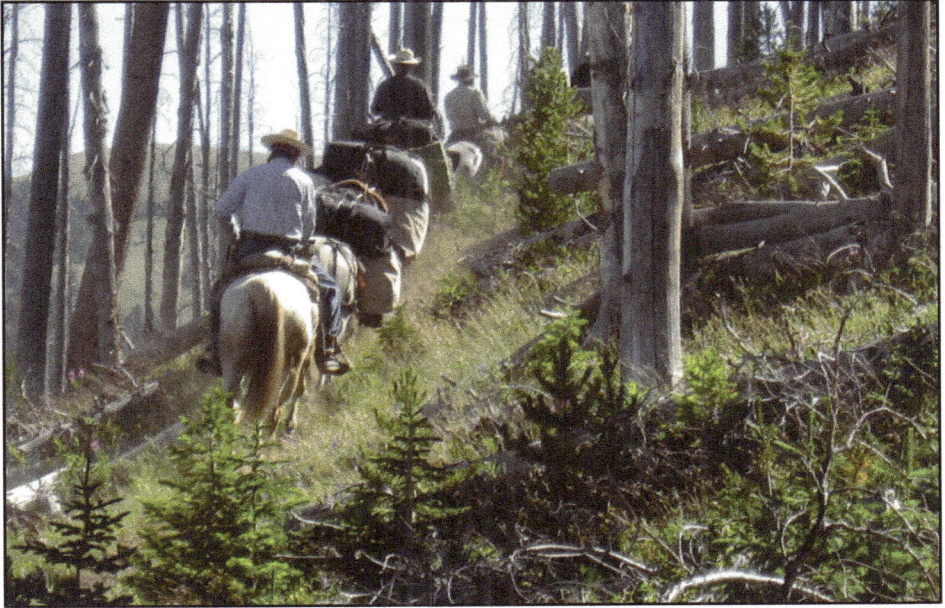

Riders retracing the 1,170-mile Flight of the Nez Perce in 2010. The men rode horses that were descended from Nez Perce stock that made the trek in 1877. **Author photograph**

For thousands of years, the stories of Native Americans and bison have been intertwined. Today, bison are relocated from Yellowstone to Tribal lands to begin or augment herds. **Neal Herbert, NPS**

Yellowstone personnel killed off the park's cougars by 1926. NPS

Chief Ranger Sam Woodring with ten captured wolf pups in 1922, four years before the park had eliminated all of its wolves. As part of its predator removal policy, the park shot and killed the pups' mother on the Blacktail Plateau. The pups were then taken to Mammoth Hot Springs as a tourist attraction for several days before park staff killed them, too. NPS

Ranger Ted Ogsten and Chief Ranger Sam Woodring with coyote pelts in 1927. Although park staff and settlers throughout the West actively killed predators, coyotes actually spread their range in the twentieth century. **NPS**

For much of its first century, Yellowstone fed its black and grizzly bears for the amusement of tourists. The sign in this photograph reads "Lunch Counter—For Bears Only" and was taken in the Old Faithful area in the 1920s or '30s. **NPS**

Bison at a Mammoth Hot Springs corral in the early 1900s. These bison were part of the park's successful restoration efforts beginning in 1902. **NPS**

Elk being driven by a helicopter in the 1960s, a time when Yellowstone's managers shot many of the park's elk. **WJ Barmore, NPS**

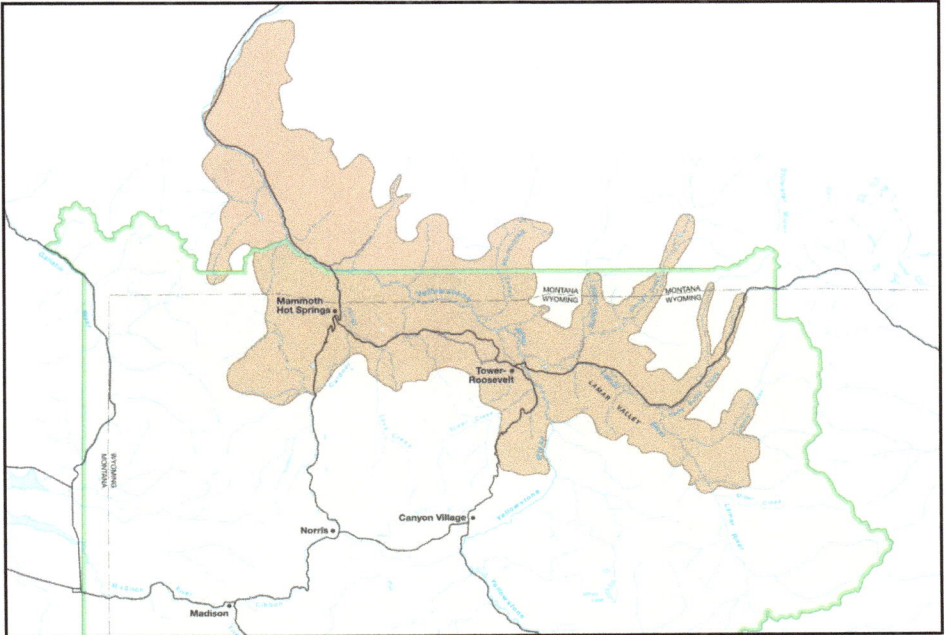

The Yellowstone area's Northern Range is home to one of the largest elk herds in the region, many of its bison and a vast assortment of other animals. **Adapted from NPS**

Pronghorn antelope, which once numbered in the tens of millions in the West, live among elk and bison in Yellowstone's Northern Range. **Author photograph**

In the mid-1990s, wolves were restored to Yellowstone after a nearly 70-year absence.
Jim Peaco, NPS

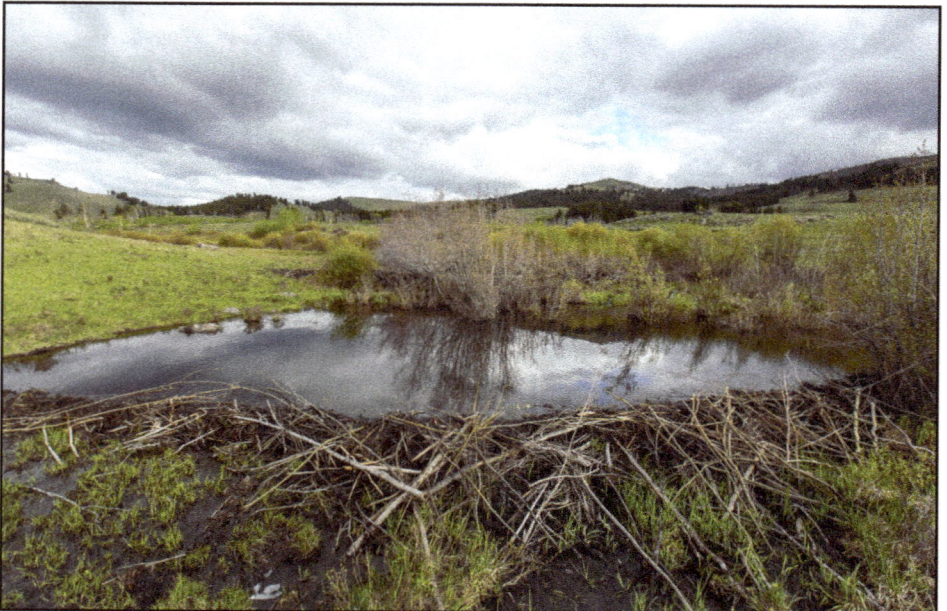

Since being restored, wolves have reduced the elk population to more natural levels on Yellowstone's Northern Range. In turn, this has led to a resurgence in willows in some areas, which has allowed beavers to return, including to the Lamar Valley, the site of this beaver pond. Neal Herbert, NPS

Cougars quietly returned to Yellowstone on their own before the restoration of wolves.
Connor Meyer, NPS

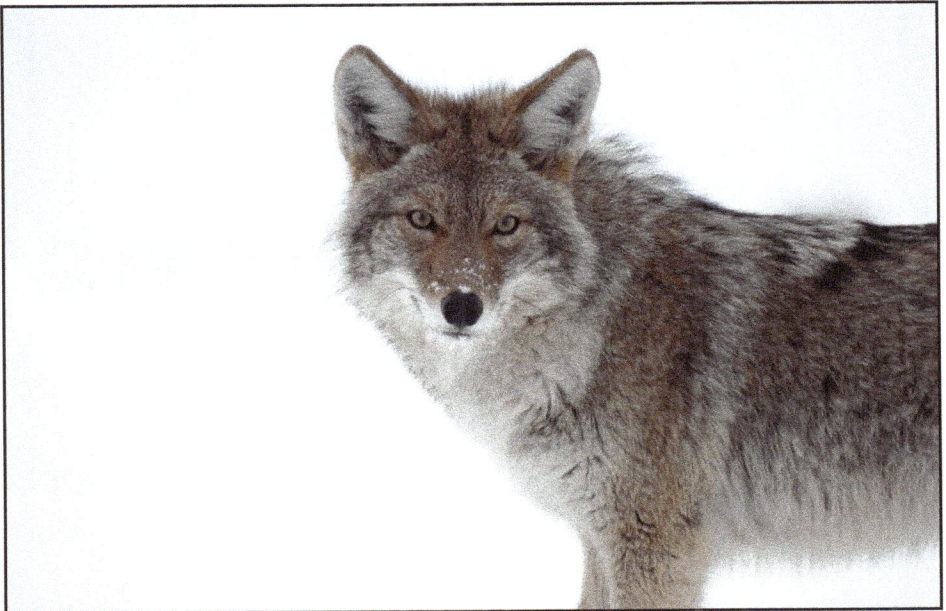

In the absence of wolves, the coyote population increased in Yellowstone during the latter twentieth century. With the restoration of wolves, coyote numbers have declined, likely to historical levels. Author photograph

Grizzly bears often take over elk and bison carcasses from wolves. Jim **Peaco, NPS**

About 100 wolves live in Yellowstone today. Jim **Peaco, NPS**

LOWER HALF OF OPPOSITE PAGE: Map showing the projected thickness of ashfall across North America should Yellowstone's supervolcano explode again. **Adapted from USGS**

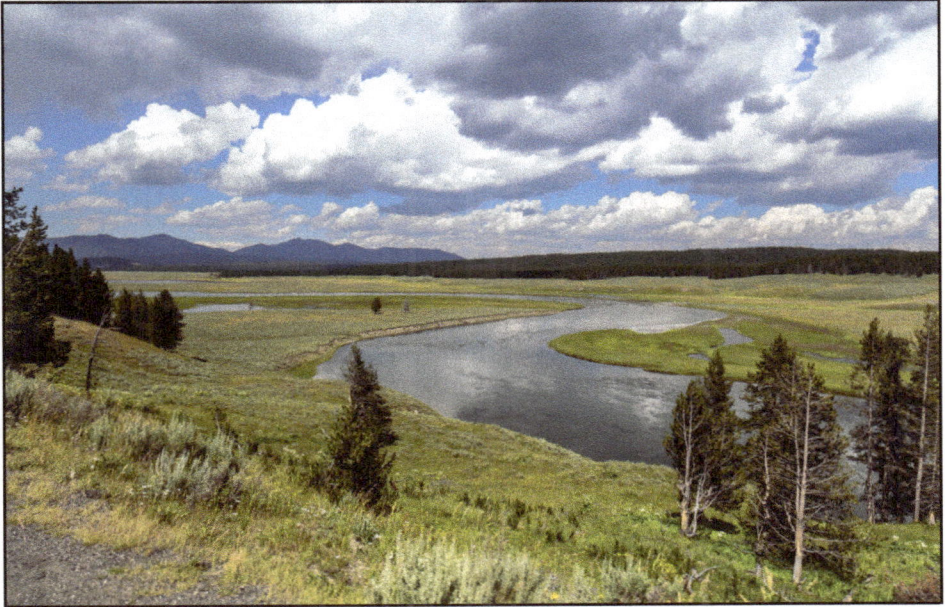

Yellowstone's Hayden Valley sits inside the northeast area of the park's youngest caldera. A new explosion would obliterate the valley and much of the park. Mount Washburn, which lost its southern half to the last super eruption, is one of the peaks in the background. **Jacob W. Frank, NPS**

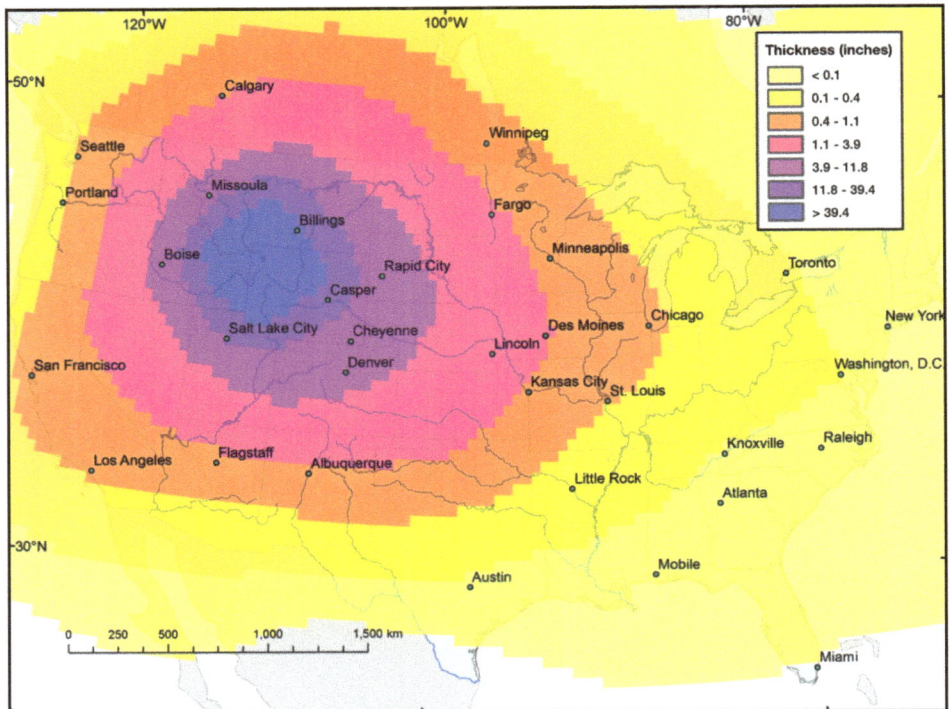

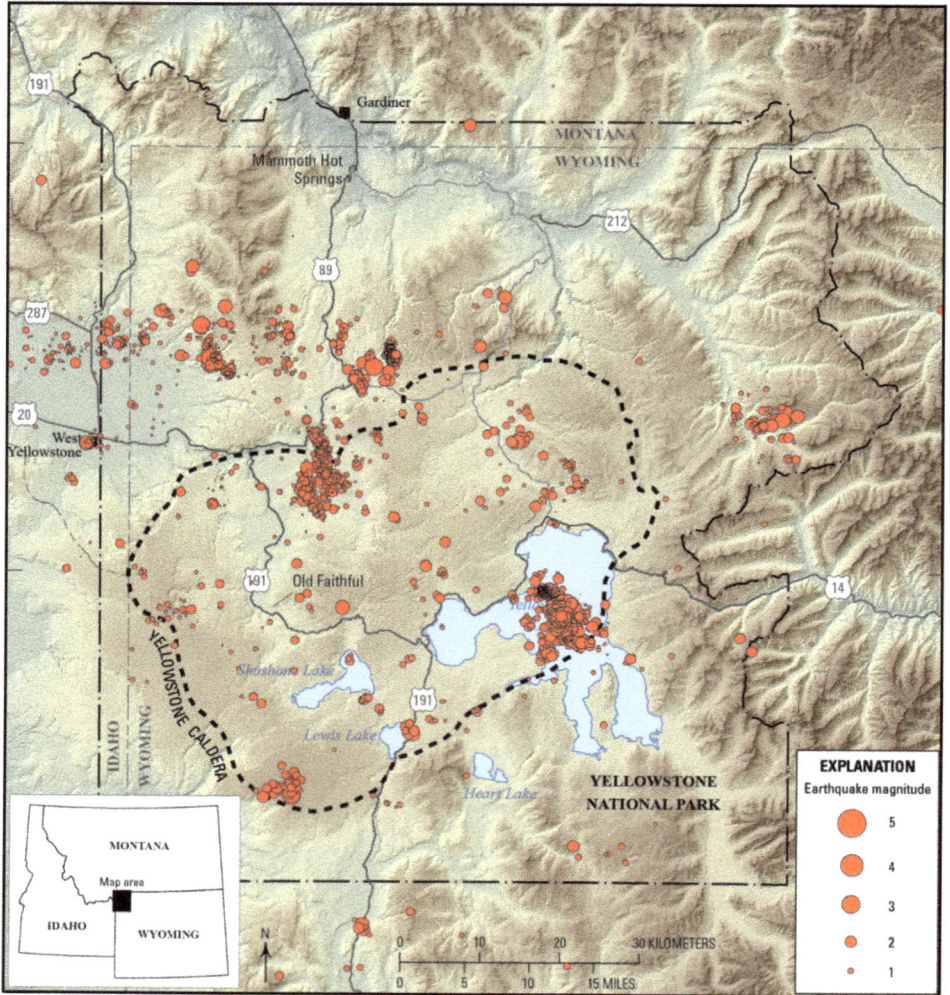

Map showing locations of the Yellowstone region's seismic activity (red circles) from 2021. Hundreds to thousands of minor earthquakes occur in the area each year. USGS

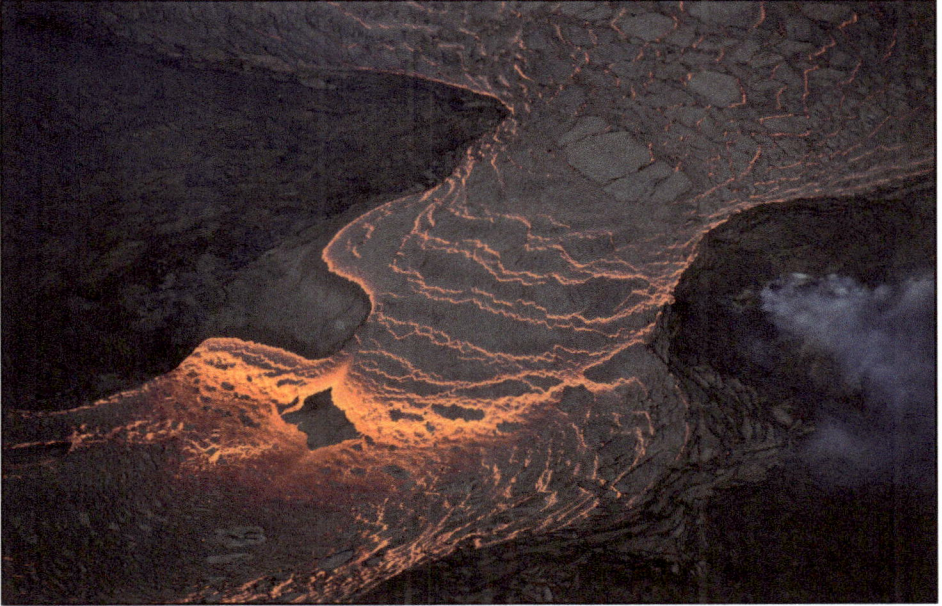

Lava flowing from Hawaii's Kilauea volcano. Lava flows from the Yellowstone Volcano are a potential danger but not likely to occur for thousands of years. **B. Carr, USGS**

A hydrothermal explosion at Yellowstone's Biscuit Basin in 2009. These types of events are the most common danger from the supervolcano, although rarely seen, let alone caught on camera. **USGS**

Photograph showing the 2016 Maple Fire burning in Yellowstone, including in areas that burned in 1988. Charissa Reid, NPS

The heat of the Maple Fire opened the waxy seal of this lodgepole pine's serotinous cone and scattered its seeds. Jennifer Jerrett, NPS

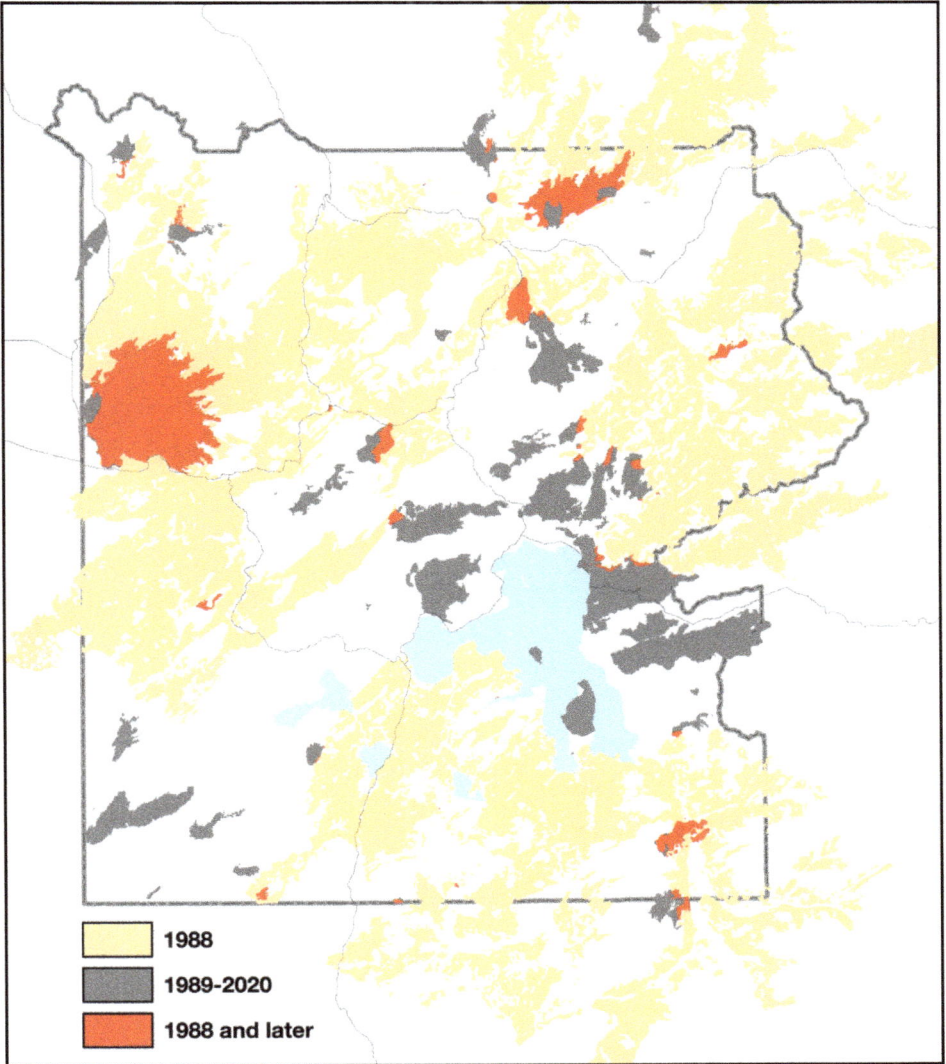

Map of burned areas in Yellowstone from 1988 to 2020. Until 2016, larger fires in the 2000s burned in areas largely unaffected by the '88 fires. In 2016, however, more than 42,000 acres burned that had also burned in 1988. NPS

Pika, which are already declining in lower altitudes across the West, might also be affected by the warming climate in Yellowstone's higher elevations. Jacob W. Frank, NPS

Animals adapted to living in forests, including pine martens, could also decline in Yellowstone should fires significantly reduce their habitat. Jacob W. Frank, NPS

CHAPTER 8

Native Americans and Yellowstone

W hen glaciers last covered much of northern North America, cold-adapted humans migrated onto the continent, beginning at least 15,000 years ago. They traveled from the mostly ice-free Bering Land Bridge that connected Siberia to Alaska. For centuries before this, they lived on the vast grasslands of Beringia and hunted mammoths, bison, elk, camels, horses and other animals while living amidst American lions, scimitar cats, short-faced bears, grizzly bears, wolves and other carnivores.

By 13,000 years ago, the first people reached the Yellowstone area as the ice melted from the landscape. They either followed an ice-free corridor from Alaska and the Yukon down into southern Montana or made their way eastward along river valleys from the Pacific Coast. Whatever route they took, they reached the Yellowstone region about the same time that the ice-age megafauna returned to the greening landscape in and around the present-day national park.

These earliest people, the Clovis Culture, lived in small family groups and moved from one area to another, depending on available food sources. They followed herds of now extinct bison, mammoths, horses, camels and other prey, hunting them with razor-sharp spears that penetrated thick fur and skin. Hunting with spears meant getting close enough to prey for lethal strikes and risking getting kicked, trampled or gored by horns, antlers or tusks. To avoid those dangers, hunters sometimes drove animals into canyons and speared them safely from above. Mammoths weighed up to 15,000 pounds, as much as the largest elephants today, and yielded as much as 5,000 pounds of meat for a family group, as well as hides for clothing and other uses.

From early on, people traveled into Yellowstone to hunt and to gather plants, and for more than 11,000 years they and their descendants quarried obsidian at Obsidian Cliff along the present-day Grand Loop Road that connects Mammoth Hot Springs to Norris. Although other sources of obsidian exist in Yellowstone and the surrounding volcanic fields, the quality and abundance of Obsidian Cliff obsidian attracted people from throughout the region and became important for everyday life and for trade. The obsidian quarried at this site has more silica content than other lava flows, making it tougher and allowing for extremely sharp edges. People used it for spearheads, to cut and process plants, to butcher animals and to scrape hides for shelter, blankets, clothing and other uses.

In the Yellowstone area, obsidian from Obsidian Cliff remained the most important source of tool-making material for thousands of years. Yet through millennia, Native Americans and their ancestors, including the Clovis people, also used bone, antlers, and ivory from animals to make tools, and they used other rock with high silica content, as well, including chalcedony, quartzite, petrified wood and chert. Some of the most highly prized chert was quarried from Yellowstone's Crescent Hill, not far from the park's Tower Junction area.

By at least 9,500 years ago, people began living seasonally in Yellowstone, especially along the shore of Yellowstone Lake in places such as Fishing Bridge, near the outlet of the Yellowstone River. They followed several routes each spring to reach the lake and moved back each winter to the valleys, including those formed by the Yellowstone, Snake, Madison and Shoshone Rivers. Animals, such as bison, pronghorn and elk, similarly migrated into the high country in warmer months and returned to the valleys for winter, as they still do today. While camping in meadows by the lake or alongside streams and rivers, people hunted various animals, from birds and rabbits to deer, elk, bear and bison. From early on, especially as other ice-age megafauna died out, bison became an important animal for Native American peoples, including two much larger and now extinct species of bison, the last of which died out by 5,000 years ago.

These people also gathered hundreds of different kinds of plants for food, as well as for spiritual and medicinal uses. The bulbs of blue camas and other roots were important foods, which people dug out with t-shaped digging sticks that featured fire-hardened tips. Antler or other cross-piece handles were attached with sinew. Burning sage and sweetgrass became important for use in sweat lodges for cleansing body and spirit, dried broomweed was used for soothing sore muscles and facilitating easier childbirth, and lichen was used for treating stomachaches, headaches and indigestion.

As centuries went by, tool-making became more sophisticated. By 10,000 years ago, people began crafting atlatls, hand-held spear throwers that extended the reach and power of their arms, allowing them to hunt more safely and successfully from farther away. Hunters chipped grooves into the bases of the spearheads and wedged them into the slotted ends of wooden shafts. Later, they began notching spearheads near the bottom and attaching them more securely to the shafts with sinew.

By 1,500 years ago, people in the Yellowstone area began using bows and arrows. Arrows used much smaller stone points and shafts than spears and also allowed people to hunt more safely from even farther away. People crafted arrowheads with stone from Obsidian Cliff, which they traveled to from throughout the region, including their high-elevation camps around Yellowstone Lake. Native peoples often quarried chunks of obsidian, roughed out blanks, and then took them back to their camps to chip into arrowheads and other tools. Blocks of Yellowstone-sourced obsidian and finished tools traveled and traded widely in all directions. Nearly 300 pounds of this obsidian was unearthed at a Hopewell Mounds burial site in Ohio and dates back 2,000 years.

As time went on, the peoples in the Americas assimilated into tribes and bands within those tribes, including in the Yellowstone area. Regular trade routes became established, and Native Americans in the Yellowstone region adopted a universal sign language spoken by dozens of Native nations to facilitate trade and communication.

People from as many as 10 different tribes lived along Yellowstone Lake during the warmer months, and more than two dozen tribes have some connection to Yellowstone, going back hundreds to thousands of years.

The roots of the Shoshone people may reach deeper than any other tribe, and more cultural sites exist in the park for them than for any other group. The Shoshone's direct ancestors may have been in the area for more than 9,000 years. Among other places, they lived in the southern area of Yellowstone and in the Absaroka and Wind River Ranges beyond the park's border. They're also known for their Intermountain Ware—pottery made from clay and strengthened with crushed rock that they used for cooking and storing food.

Like other tribes, the Shoshone saw Yellowstone as a spiritual land. Individuals undertook lengthy fasts, often on mountain tops, in and near Yellowstone. From their stone fasting beds, they sought knowledge and strength from the spirit world. They also soaked in hot springs to bathe and to soothe muscles, arthritis and other ailments.

When whitebark pine nuts matured in the fall, they traveled to the high country in the Absaroka and Wind River Ranges and set up extensive camps to harvest the protein- and fat-rich nuts, storing them for winter use. Some such villages may date back 4,500 years and some had been used continuously from 1,000 years ago and into the nineteenth century. While in the high country, the Shoshone also hunted bighorn sheep.

The Blackfeet Nation also harvested whitebark pine nuts, though farther north along the border of present-day Montana and Canada. They sometimes followed Clark's nutcrackers to find trees with abundant cones, which they knocked to the ground with long sticks. They charred the cones in campfires to loosen the nuts and then roasted them. It wasn't until the Blackfeet adopted the use of horses and rifles in the 1700s that they ranged farther south, making trips to hunt bison and other animals on the Great Plains and in the Yellowstone area, including to the northern reaches of Yellowstone Lake.

The Crow people (Apsáalooke) also spent some of their time in the northern part of Yellowstone as well as the east. They met the Shoshone and Nez Perce (Nimiipuu) each summer for trade, games and celebration at the headwaters of the Yellowstone River; and they, like the Shoshone, also trekked to thermal areas for their healing waters, as well as to fast and to seek visions.

While the Crow, Blackfeet and Shoshone moved in and out of Yellowstone seasonally, the Sheep Eaters (Tukudika) band of the Shoshone might have been the only Native Americans to have lived year-round in the park, perhaps for as long as 3,500 years. They generally followed migrating bighorn sheep, hunting them for their meat, horns and hides. They softened the horns in hot springs and fashioned bows out of them. Other tribes traded for these bows as well as for clothing, hides and obsidian. In return, the Sheep Eaters sought bison robes, salmon and other items.

The Sheep Eaters also fished and gathered roots, seeds, berries and pine nuts. Some of the roots, such as camas and balsamroot, could be eaten raw or baked, and people also dried them for use in soups or ground them into flour.

Unlike other tribes in the area, the Sheep Eaters living in what's now Yellowstone didn't use horses. Yet they did rely on dogs for hunting, for guarding camps and for pulling travois: sledges consisting of two poles that fanned out from a dog's harness or collar, with a net strung between them. The dogs dragged travois loaded with belongings behind them as their people moved from area to area.

Osborne Russell, a fur trapper and one of the first European Americans to step foot in Yellowstone, met a Sheep Eater clan in the Lamar Valley in 1834. He counted 30 dogs living with about two dozen people. And he wrote of the beauty of their bows made from sheep and bison horns, as well as from elk antlers, and he remarked on their well-made buckskin clothing.

To hunt sheep, the Sheep Eaters sometimes used dogs to drive them into deep snow where they could be easily killed. They also built wooden drives to funnel sheep into traps where they also could be easily dispatched.

The Bannock, another Shoshone band, didn't live in Yellowstone year-round like the Sheep Eaters, but they would have traveled by Sheep Eater camps on their seasonal migrations from eastern Idaho's Snake River Plain through Yellowstone and onto the Great Plains to hunt bison. The Bannock entered what's now the park near West Yellowstone and traveled roughly northeast toward present-day Mammoth Hot Springs before heading east toward the Lamar Valley and beyond. Along this route, known as the Bannock Trail, parts of which can still be seen today, other trails led north to more hunting grounds along valleys to the west of the Great Plains.

Not long after Osborne Russell encountered a Sheep Eater family, the ancient connection between Native Americans and Yellowstone withered. This was about the same time that Yellowstone became a national park in 1872 and began welcoming tourists while shunning indigenous peoples.

By that time, European Americans had been moving steadily westward, taking over Native homelands and decimating the wildlife that sustained tribes. In the eastern United States, skirmishes and battles over land and resources between European settlers and Native Americans began shortly after colonists arrived in Jamestown, Virginia, in 1607. After more and more settlers moved onto Native homelands throughout the next 220 years, President Andrew Jackson signed into law the Indian Removal Act of 1830, allowing the US government to force Natives living east of the Mississippi River from their homes.

Eight years after Jackson signed the act, soldiers descended on Cherokee Nation lands in Tennessee, Georgia, Alabama and North Carolina, forcing 16,000 of their people to walk 1,200 miles west to Indian Territory (present-day Oklahoma). At least 4,000 died from disease and starvation along the way in what became known as the Trail of Tears. Other forced removals occurred in the Southeast during the 1830s with other tribes, including the Creek, Chickasaw, Choctaw and Seminole. In all, as many as 100,000 people were forcibly removed from their homelands and marched to Oklahoma.

While this took place, Native Americans in the Yellowstone area continued to live largely as they had for centuries. They came into contact only occasionally with white people, including explorers and fur trappers like Osborne Russell, although smallpox and other diseases traveled along trade routes and took their toll on the tribes.

The Nez Perce, like the Bannock, historically traveled through Yellowstone from their ancestral lands to the west to hunt bison on the Great Plains and elsewhere. And they, like other tribes, used Yellowstone's thermal features for ceremonial purposes. In certain places, lodge poles from their shelters from the 1800s still remain standing or lay scattered in the park. The Nez Perce built these shelters (called "wickiups") from nearby stands of lodgepole pine, which grow thickly in Yellowstone and neighboring forests. Lodgepole pine gets its name because it generally grows straight, making it ideal to frame shelters.

Five years after the establishment of Yellowstone National Park, the Nez Perce would make their last major encampments there. One of these camps, high above the surrounding valleys, gave the tribe a commanding view of the landscape below. In the past, from this spot near the headwaters of the Lamar River, they scanned the valleys and hills below for prey and for other tribes, but this time in 1877, they watched for the pursuing US Army.

In late spring of that year, the Nez Perce fled their homeland after a relatively quick series of events. In 1855 as waves of settlers moved onto their lands, the Nez Perce signed a treaty ceding more than half of their traditional homeland—about 17 million acres in Washington, Oregon and Idaho—to the US government. The treaty set aside 7.5 million acres as their reservation and permitted them use of the ceded land to hunt and fish.

A year after the Senate ratified the treaty in 1859, trespassing settlers found gold on the new reservation, bringing thousands more streaming in to prospect and to claim land. The federal government did nothing to stop this, and in fact, forced a new treaty in 1863 that shrank the reservation by 90 percent, to 750,000 acres. The new treaty divided the Nez Perce. Generally, those bands of the tribe living within the proposed new boundaries agreed to it while those living beyond it walked out of talks.

For 10 years after the Senate ratified the second treaty in 1867, the non-treaty Nez Perce lived a tenuous existence. In the spring of 1877, the government ordered them to move onto the reservation. The Nez Perce grudgingly appeared ready to do this, but in June of that year, several warriors, seeking revenge for all they had lost, raided white settlements, killing more than a dozen people. With no turning back, about 800 men, women and children along with nearly 2,000 horses fled.

The army pursued the Nez Perce with orders to arrest those who had killed the settlers and to force the others onto the reservation. On June 17 at White Bird Battlefield in western Idaho, the army attempted to carry out their orders but met fierce fighting and were routed, losing 34 men while the Nez Perce lost none. White Bird became the first of six battles between the Nez Perce and the army, along a 126-day, 1,170-mile trek, and it occurred 14 months after the

Lakota Sioux, the Northern Cheyenne and the Northern Arapaho tribes had wiped out Lt. Col. George A. Custer and his men at the Battle of Little Bighorn.

After more than two months' flight through Idaho and Montana, the Nez Perce arrived in Yellowstone in late August of 1877. They spent nearly two weeks in the park, entering it by present-day West Yellowstone, riding to the Lower Geyser Basin and then crossing the center of the park into Hayden Valley. South of there, they forded the Yellowstone River and then followed the north shore of Yellowstone Lake before ascending to their camp near the Lamar's headwaters.

After striking that camp, they descended from the Absaroka Mountains onto the plains in Wyoming and then Montana, where they sought help from their old friends, the Crow. Rebuffed by that tribe, they turned north hoping to find safety in Canada, where Sitting Bull and other Sioux had gone.

Just 40 miles from the border and with cold weather settling in, the army caught up with the Nez Perce. For six days, the two sides fought as the army laid siege to the tribe's camp. After being promised that his people would be allowed to return to the reservation in Idaho, one of the tribe's leaders, Chief Joseph, agreed to surrender on October 5. Not all Nez Perce agreed with his decision, however, and in the dark of night, about 30 of them escaped into Canada to join several hundred others who had previously crossed the border.

Native Americans and Yellowstone

The more than 400 Nez Perce who stayed in Montana at the Bear Paw Battlefield soon learned that the government had gone back on its word. Instead of allowing the tribe to return to Idaho, the army forced them to march more than 250 miles to southeast Montana, and eventually sent them by rail first to Fort Leavenworth, Kansas, and then to Oklahoma, where the government had forced the Cherokee to march nearly 40 years earlier.

A year after the Nez Perce traveled through Yellowstone, a smaller group of Bannock and Paiutes fleeing the army followed part of their route in the hopes of reaching Canada. The army surprised them, however, just beyond the park's border and either killed or captured them. This subset of a larger group that had rebelled months earlier to the west of Yellowstone had done so in part because of hunger on the Bannock reservation at Fort Hall in Idaho. Despite a treaty with the US government, settlers had turned their pigs loose onto the camas meadows that the Bannock traditionally foraged in, causing extensive damage. After the uprising ended, some Bannock and Shoshone made forays from the reservation to hunt inside the park into the 1890s even as the US Army began more regular patrols from Fort Yellowstone in Mammoth Hot Springs.

In 1916, the newly created National Park Service took over day-to-day operations of Yellowstone from the army, and by that time, the area's Native Americans had largely settled onto reservations or in towns and cities in Montana, Wyoming and Idaho. Not until later in the twentieth century did indigenous people begin to have a say in how Yellowstone should be managed, rekindling a presence that dated back millennia.

Today, members from numerous tribes advise park managers about access to and the importance of cultural resources in Yellowstone. This includes the collection of plants and obsidian for ceremonies, hunting near the border of the park and the proper reburial of ancient human remains found in Yellowstone. They also work on achieving more diversity among park employees and interns, and they seek to teach the millions of tourists visiting the park each year about their long connection to Yellowstone and its surrounding lands. And they play an important role in securing the future of bison, an animal whose destiny has been linked to theirs for thousands of years, and which nearly disappeared by 1900.

CHAPTER 9

The Decimation and Restoration of Wildlife

lthough a majority of North America's large mammal species died out by 10,000 years ago, after the arrival of hunter-gatherers, the wildlife that persisted—including bison, elk, bighorn sheep, pronghorn, mule deer, grizzly bears, wolves and cougars—adapted and coexisted with them.

The arrival of European Americans to the Yellowstone area threatened this thousands-years-old balance, however, beginning in the mid-1800s. That balance had already shattered across much of the East by this time, as Europeans began settling along the Atlantic Coast in the 1600s and moving steadily westward after that. These immigrants and their descendants cleared forests and plowed under vast swaths of prairie and other land to make room for crops, livestock, towns and cities. As they altered these landscapes, they killed off much of the wildlife living in them.

By the time they began settling in the valleys around Yellowstone in the mid-1800s, European Americans had all but eliminated elk, bison, cougars and wolves throughout the eastern United States. They then set their sights on them in the West.

In the early to mid-1800s, tens of millions of bison ranged throughout the Great Plains and Western states, but by the time Yellowstone became a national park in 1872, only a fraction of this number survived. And by 1900, just 23 of these wild bison remained, in Yellowstone's Pelican Valley.

Hide hunters slaughtered the great herds for their valuable skins, leaving millions of carcasses to rot, although some of the meat fed railroad workers, settlers and soldiers. The US Army also deliberately reduced the herds to take away an important food source of Native Americans, thereby forcing them onto reservations. Other people gathered bison bones for profit, sending them off by rail to become fertilizer and bone china.

Like bison, elk suffered greatly as well. Before European settlement, an estimated 10 million lived throughout much of what's now the United States, but by 1900, their numbers had dwindled to 41,000. In Montana, where they once ranged throughout much of the state, they existed only in and near Yellowstone and along some areas of the Flathead River far to the north-northwest of the park.

Even after the United States designated Yellowstone a national park in 1872, hunters slew large numbers of elk inside its borders. In 1875, the Bottlers, who ranched 30 miles north of the park near Emigrant, Montana, slaughtered some 2,000 elk for their hides inside and outside of Yellowstone's northern border. That area today provides habitat for up to four times that number of elk; and had hunting not been outlawed inside the park in 1883 and poaching strictly enforced starting in 1900, elk could have vanished there, too.

An estimated 35 million pronghorn also once ranged throughout the American West, but by the early 1900s, fewer than 15,000 survived. In 1874, 12,000 pronghorn hides were shipped by rail just from Bozeman, Montana, alone.

Bighorn sheep numbers also dropped dramatically as European Americans settled in the Rocky Mountains. People slaughtered them for their horns and meat while domestic sheep spread fatal diseases to them.

Greatly overhunted, both mule deer and white-tailed deer herds also suffered during this time. Mule deer, which once ranged throughout Montana, nearly vanished from many areas in the state.

As the great herds of ungulates disappeared, so, too, did the large predators that depended on them, including wolves and cougars, which had turned more and more to preying on livestock to survive. Settlers shot, trapped and poisoned them, as well as coyotes, black bears and grizzlies—and any other wildlife they saw as a threat to their livestock, crops or safety.

As official policy, Yellowstone National Park personnel also killed many of its predators beginning in the late 1800s and continuing until 1934. By 1926, no wolves survived in the park, and the more secretive cougars likely vanished, too, the last known one being killed in 1925. Coyotes, although persecuted inside and outside the park, actually expanded their range from the West and Midwest to nearly all of North and Central America by the late 1900s, taking advantage of the absence of larger carnivores and the fragmentation of forests by people.

Although killed in the lands around Yellowstone, grizzly and black bears became entertainment for tourists inside the park, beginning in the 1880s and lasting until 1970. During that time, visitors fed bears by the roadsides and watched them foraging for garbage in the park's dumps. Although cougars, wolves and coyotes were routinely targeted, Yellowstone didn't have a policy to kill off bears. Because of that, grizzly bears, like elk and bison, found refuge in the park while their overall numbers in the Lower 48 dropped sharply—from a high of about 50,000 to about 1,500 today. They exist in just 5 percent of their historic range in the Lower 48, in pockets of Montana, Wyoming, Idaho and Washington. No longer do they wander in Oregon, California, Colorado, Utah, Nevada, Arizona, New Mexico, Oklahoma, Texas, Arkansas, North Dakota, South Dakota, Nebraska, Kansas, Minnesota, Iowa and Missouri.

With the absence of wolves and cougars in the ecosystem, along with fewer grizzlies, the elk population steadily rose higher than it probably ever had. Protection from poaching and hunting inside Yellowstone, combined with regulated hunting outside of it, also contributed to the increase in elk.

One of the largest elk herds in the ecosystem lives on the Northern Range. The extensive grasslands that make up this range surround the Yellowstone and Lamar Rivers and their tributaries and provide habitat to one of the largest and most diverse assemblages of wild animals on Earth. About 80 percent of the Northern Range lies in the northernmost parts of Yellowstone while the rest reaches beyond the park's borders into Montana. With lower elevations and less snowfall than other places in Yellowstone Country, this range offers easier foraging for elk, bison, deer and pronghorn. Many of these animals migrate from the park in winter, and the more severe the winter, the greater the number of animals that move out.

By 1910, the State of Montana began trapping some elk in winter after they left the park. It shipped these throughout Montana and to several other states to augment existing herds or to start new ones. In all, more than 7,000 Northern Range elk were shipped by rail or truck during a nearly 60-year period. Additionally, Yellowstone officials trapped and shipped more than 13,500 elk from the early 1930s through 1967 to build up or restore herds in 38 states, as well as in Canada and Mexico.

As elk numbers climbed in the first half of the twentieth century, bison began a more limited revival, after having nearly vanished by 1900. In 1902, 21 bison from private herds in Montana and Texas arrived in Yellowstone to bolster the small herd from Pelican Valley. The park kept these bison (18 pregnant cows and 3 bulls) in an enclosure at Mammoth Hot Springs along with some calves from Pelican Valley. Within five years, the herd had grown too large for the Mammoth compound, so park managers moved them to the Lamar Buffalo Ranch along Rose Creek in the Lamar Valley. For several decades, bison moved freely through the valley in summer months while managers herded them back to corrals for winter, feeding them hay grown on the 600-acre ranch.

Within about 30 years, bison, which have a high reproductive rate, increased from fewer than four dozen animals to more than 1,000 between the Lamar and Pelican Valleys. In 1936, managers seeded new herds in central areas of the park, releasing 35 bison in the Firehole Valley and another 36 in Hayden Valley. Yellowstone also shipped bison out of the park to stock herds on tribal and public lands. In the early 1950s, the park stopped feeding bison at the Buffalo Ranch and closed that facility, as well. Bison roamed the park freely from that point on.

As elk and bison numbers grew, managers became concerned about them overgrazing the Northern Range. They not only continued to trap and relocate some of them but began killing them, too. From the early 1930s until 1968, Yellowstone officials killed nearly as many elk as it shipped out, including more than 4,000 in the winter of 1961–62. And beyond the park's borders in Montana, hunters legally killed another 45,000 in that period.

Yellowstone personnel also culled hundreds of bison during that time and reduced pronghorn numbers, as well. They mistakenly thought pronghorn over browsed sagebrush, and so they aimed to keep the herd's population at 150 or less, about a third of the average number of pronghorn in the decades before. They shipped many of the trapped pronghorn to other places to build herds while shooting others.

Even as the park and hunters removed elk, the herd not only grew but continued to browse heavily on willow, aspen and cottonwoods—so much so that by the 1950s, new growth of these trees and shrubs essentially ceased. Their decline also led to other losses.

In the early 1920s, an estimated 200 beavers lived along Elk Creek in the Tower Junction area of Yellowstone, but by the 1950s, none remained. Elk had severely browsed back the aspen and willow that the beavers needed for food and to build their dams. In all, researchers counted just eight beaver colonies on the Northern Range in 1953 compared with 25 in 1921. Other animals and nearly three dozen songbird species also likely declined with diminished habitat and fewer places to nest.

And, with no wolves keeping their numbers in check, coyotes became more abundant and likely caused declines in smaller predators and raptors who hunted similar prey, including ground squirrels, mice and voles. Coyotes also likely preyed on more pronghorn fawns, contributing to a reduction in that herd's numbers after the park stopped trapping and shooting them.

In 1969, the park stopped wildlife removals and adopted a hands-off policy allowing natural regulation to occur. Yet top predators, especially wolves, remained missing from the landscape, and elk numbers quickly climbed.

Aldo Leopold, a twentieth-century ecologist, suggested in 1944 that Yellowstone Country needed wolves and other predators to reduce prey numbers to more natural levels, thereby leading to healthier landscapes. More than a quarter century later and two years after the Endangered Species Act became law in 1973, Yellowstone began planning for the restoration of wolves to the park. Nearly 20 years later and following an extensive public engagement and commenting period, as well as setbacks and advances in Congress and the courts, the federal government gave the go-ahead to bring wolves back to Yellowstone, as well as to the central Idaho wilderness.

Working with Canadian wildlife biologists, the US Fish and Wildlife Service trapped 14 wolves east of Jasper National Park in Alberta in January of 1995. The topography of the area resembles the

Northern Range, where the wolves would eventually be released, and some of the elk these wolves had been hunting might have descended from Yellowstone elk transplanted there earlier in the century.

On two different occasions that January, officials flew wolves to Montana and then trucked them to Yellowstone in horse trailers. At three different locations on the Northern Range, the wolf reintroduction team hauled the crated wolves, a mix of adults and pups, onto horse-drawn sleighs and then carried them to one-acre pens about a mile from the road. The wolves lived in these pens for 10 weeks to get used to their new home. The team hauled meat from driver-killed deer, elk, moose and bison to the pens twice a week until they released the wolves in March. One of the pens still stands near Rose Creek in the Lamar Valley, uphill from where the park restored bison nearly a century before.

The team trapped 17 additional wolves in British Columbia in 1996, shipped them to Yellowstone and placed them in four acclimation pens for 10 weeks before releasing them. During both years, wolves bred in the pens, and by the beginning of 1997, the wolf population had grown to 51 wolves in nine packs. Their numbers continued growing through the next several years, reaching highs of more than 170 wolves in the mid-2000s and averaging about 100 since 2009. More than 400 additional wolves live outside the park in surrounding public and private lands, although wolves from both populations move freely across borders.

By the time wolves returned in 1995, cougars had quietly recolonized Yellowstone and its surrounding lands. Grizzly bears, which like wolves gained protection under the Endangered Species Act, also continued to recover in the ecosystem.

Today, with Yellowstone's top three historical predators in place, as well as with pressure from black bears and coyotes, elk numbers on the Northern Range have dropped from highs of nearly 20,000 in the years before wolf restoration to below 8,000. Adult elk are the primary prey for wolves and cougars as they were for thousands of years, while grizzlies, black bears and coyotes generally prey on elk calves in the late spring and early summer. Cougars and wolves also hunt deer; and increasingly, wolves are taking down more bison and scavenging on others that die during the rut, in winter and at other times.

Predator and prey numbers naturally fluctuate over time depending on the densities of each—along with climate, disease and other factors—but wolves played the leading role in driving down a high elk population, as well as altering elk behavior and habitat selection.

Following the reduction of elk, aspen, cottonwood and willow have resurged in many, but not all, places. A change in elk browsing behavior appears to be a key factor in this. And as willow resurges, so have beavers throughout the park, although not at Elk Creek. In the long-term absence of beavers, the water table dropped, and in a warming, drier climate, wetlands likely won't return there. Yet along other creeks and rivers where willow grows thick and tall again, beavers have moved back in along with greater numbers of warblers and other songbirds. Beaver-dammed areas also benefit many other animals, including frogs, toads, salamanders, waterfowl, moose and deer.

While reducing elk numbers, wolves also knocked the Northern Range's coyote population in half in the initial years after restoration, often killing them near carcasses when the much smaller coyotes came in to scavenge. Wolf-killed elk and bison continue to provide an important, though risky, food source for coyotes, as well as play an important part in the lives of many other animals—from fungus, flies and beetles to ravens, eagles and grizzly bears.

On average, more than two dozen ravens appear at wolf kills to feed. Magpies and golden and bald eagles often join them as wolves rest nearby. Smaller carnivores, including foxes, which likely benefit from fewer coyotes around, also scavenge kills when they can do so safely.

Aside from wolves, grizzly bears might benefit the most from wolf kills, especially when other foods are scarce. While smaller scavengers run or fly from wolves returning to their kills, more often than not grizzlies successfully take over carcasses from a pack. In at least three straight years, they stole every observed kill that the

Mollie's Pack made in Pelican Valley. And researchers have watched varying numbers of wolves and grizzlies interacting near carcasses, including 12 grizzlies and 4 wolves at one kill and 4 grizzlies and 10 wolves at another. In both cases, the largest grizzly controlled the scene. A lone grizzly also once successfully kept 24 wolves from reclaiming their kill. And grizzlies likewise seize kills from cougars.

With wolves, cougars and grizzlies thriving again, all the native predator and prey species that coexisted with Native Americans in Yellowstone for thousands of years call the park home once more. Although some species lag behind their historic numbers, including bighorn sheep and pronghorn, few of Earth's other temperate ecosystems are as intact and diverse as Yellowstone's.

Had the Yellowstone Hotspot not elevated the land and fueled the thousands of thermal features Yellowstone is famous for, thereby gaining it protection as a national park, many of the wildlife species found there today would be gone or greatly reduced. So even though the Yellowstone Volcano's explosions decimated the area long ago, its very existence helps to preserve the park and its inhabitants today. How long that lasts depends on when the supervolcano erupts again.

CHAPTER 10

The Threat of the Supervolcano

I magine a late spring morning in Yellowstone's Hayden Valley. Scattered herds of bison graze on the rich green grass; a grizzly bear searches sagebrush for hidden elk calves; geese, pelicans and ducks feed along the Yellowstone River; and near a cluster of lodgepole pines, wolf pups play in the sunshine by their den. Hundreds of tourists drive along the road or have stopped to watch the animals.

Suddenly, the land shakes violently. And then just as abruptly, lava explodes from the ground along an elliptical path tens of miles wide. The explosion obliterates the wildlife, the people, the valley, the river, Yellowstone Lake to the south—and everything in the central area of the park, including forests, geyser basins, campgrounds, hotels and visitor centers. Gone also is much of Mount Washburn, which had lost its southern half to a super eruption 640,000 years ago.

As rock, gas and ash billow more than 12 miles into the sky, lava and gas race out from the new caldera and incinerate all living things in the way. Thousands more park visitors and employees perish instantly, as does all wildlife in the area. Just the mountainous northwest, northeast and southeast regions of the park escape the pyroclastic flows.

But the cataclysm isn't contained to just Yellowstone. The enormous eruption not only drops several feet of ash onto gateway towns in Montana, including Gardiner, West Yellowstone, Silver Gate and Cooke City, but also on the communities of Billings and Bozeman in Montana, Cody and Jackson in Wyoming, and Idaho Falls, Idaho.

Virtually no western state escapes the ashfall, although accumulations lessen the farther from the volcano it settles. One to three-and-a-half feet of ash accumulates in Salt Lake City, Denver, Boise and Missoula; up to four inches fall on Spokane, Fargo, Rapid City and Lincoln; nearly an inch falls on Seattle, San Francisco, Los Angeles, Las Vegas, Albuquerque, Kansas City, Minneapolis, Chicago, Calgary and Winnipeg; and ash dusts nearly every other city and town in the contiguous United States and southern Canada.

Transportation in the Yellowstone area comes to a standstill as the ashfall thickens. At first, it grounds air travel in the region as even small amounts of ash in the atmosphere can damage jet engines. Eventually, however, the enormous ash cloud spreading outward in every direction will ground all planes in the contiguous United States and affect air traffic worldwide for months.

Ground travel also slows nearly to a halt as speeds greater than five miles per hour kick up ash, reduce visibility to near zero and lead to accidents on the slippery, ash-covered roads. Ash also clogs air filters and causes internal-combustion engines in cars and trucks to stall. Thousands of motorists are stranded on roads and highways.

Heavy layers of ash, especially when soaked by rain, knock out power lines, transformers and cell towers, shutting down electricity and phone service for weeks to months. Hospitals, nursing homes and other care facilities without back-up power face emergencies. And roofs with heavy loads of ash collapse under the weight of it, burying and trapping people. First responders and search-and-rescue crews will have many days of hard work ahead of them.

Volcanic ash isn't light and fluffy like the ash from burning forests, but is comprised of tiny, jagged shards of extremely abrasive rock and natural glass. It doesn't dissolve in water and can stick around for a long time. Exposure to it irritates eyes and skin and causes shortness of breath, coughing and other ailments, especially for children, the elderly and those with already compromised respiratory and cardiovascular conditions, including asthma and emphysema.

Ash particles easily creep into cracks in homes and other

buildings and damage computers, electronics, appliances and other machinery. It also contaminates some communities' drinking water and damages or destroys crops growing close to the Yellowstone area. Grazing animals—wild and domesticated—suffer from inhaling and ingesting ash as they eat. Countless numbers will eventually die.

Even after the ash has fallen, wind and other disturbances will kick it up, continuing to harm human health and society, as well as all living things for months to come. In the first few weeks, millions of people are injured or killed by the eruption, ash inhalation, accidents and collapsed homes and buildings. It's the greatest natural disaster ever to hit the United States. But the effects of the super eruption also have global consequences, the scale of which humanity has never faced before.

Besides ash, the Yellowstone Volcano expelled copious volumes of gases into the atmosphere. One of these, sulfur dioxide, forms sulfate aerosols in the stratosphere that reflect sunlight back into space. These aerosols will cause global cooling for several years that greatly impacts food production, power generation and global travel and trade.

On a much smaller scale, the global temperature dropped nearly one degree Fahrenheit following the eruption of Mount Pinatubo in the Philippines in 1991. And 1816 became known as "the year without a summer" following the even-larger 1815 eruption of

Tambora in Indonesia. Many crops around the world failed in 1816 and led to many more deaths from starvation than the death toll of 70,000 in the immediate aftermath of Tambora's eruption.

Yet Tambora ejected only about a seventh of the ash and rock as the most recent Yellowstone explosion. A massive eruption of the Yellowstone Volcano would lead to a much harsher volcanic winter. Cold, snowy weather might persist nearly year-round in higher latitudes, making agriculture there nearly impossible, and erratic rainfall in more tropical areas could result in drought or flooding and lead to large-scale crop failures. Many millions of people would die of starvation worldwide.

Although a caldera-forming eruption of the Yellowstone Volcano would be horrific, in truth, we wouldn't be caught off guard and would likely have enough time for mass evacuations and disaster preparedness.

Scientists would detect a pending explosion weeks or months ahead of time, and in all likelihood, such an eruption is thousands of years off and might actually never happen again in Yellowstone. By the time the Yellowstone Hotspot's magma chamber is primed to explode again, it may be a hundred miles or more northeast of the park where the rising magma might fail to breach the thicker continental crust that lies beyond Yellowstone.

Each year, research yields new information about the Yellowstone Volcano, one of the best-monitored volcanos in the world. By analyzing the seismic waves from the one to three thousand earthquakes that rattle Yellowstone each year, scientists from the University of Utah have mapped the upper magma chamber. At about 145 cubic miles, the amount of magma is much less than the volume of material ejected by the largest of the super eruptions 2.1 million and 640,000 years ago. The oldest eruption ejected 600 cubic miles while the younger one shot out 240. Seismic waves, which travel faster through solid rock than molten rock, also reveal that only about 10 to 15 percent of the chamber is molten. And that molten magma sits in pockets here and there amongst denser rock. To erupt, molten magma needs to increase, pool together and become highly pressurized.

Further, the lower magma chamber, which is more than four times as large as the upper chamber and could fill the Grand Canyon 11 times over, is thought to be only about 2 percent molten. To feed an eruption, the lower chamber also needs to become more molten and pass some of that magma to the upper chamber. And this rise of magma is what the network of monitoring devices around Yellowstone Country would identify.

Numerous seismometers continuously detect earthquakes and earthquake swarms in and around Yellowstone. Swarms can last from days to months and consist of hundreds to thousands of smaller quakes. The magnitude and frequency of these quakes would greatly increase as the magma rises deep below. GPS and other satellite-based monitoring would also detect ground levels rising significantly higher than previous measurements, as much as two feet or more per year, as magma inflates the land above it. And regular monitoring of stream flows, hydrothermal activity and gas emissions also help scientists detect changes in the volcano.

Should earthquake activity increase and grow stronger and the ground rise rapidly during the course of months or years, scientists would have enough time to warn people. They knew ahead of time that Mount St. Helens was ramping up before it blew in 1980. For two months, more than 10,000 earthquakes preceded that eruption, and the north flank of the volcano bulged outward by more than 250 feet.

Yet in Yellowstone, any increased earthquake activity and ground deformation might very well calm down for centuries or millennia. The quakes might lessen and the ground could subside as magma and hydrothermal fluids drain from the upper chamber without an eruption at the surface. This ramping up and settling down has been happening in Yellowstone for thousands of years.

Even if Yellowstone's volcanic system was priming for an eruption, it's much more likely that it would be a non-explosive lava flow with a minimal amount of ashfall and rockfall outside of the park. A lava flow, though potentially massive, would move slowly and have only local consequences. Eighty such flows have occurred since the last super eruption. Some are the largest ever known on

Earth, flowing nearly 20 miles from their vents and thickening to as much as 1,300 feet. The last occurred about 70,000 years ago and formed the Pitchstone Plateau in the southwest of the park.

Lava flows associated with the Yellowstone Hotspot include basaltic and rhyolitic ones and can last anywhere from a few weeks to years. The less viscous basaltic lava erupts more freely and rapidly than rhyolitic lava. Basaltic flows occur more frequently along Idaho's eastern Snake River Plain, the area the hotspot previously influenced. In the past several thousand years, lava has flowed in places such as Idaho's Craters of the Moon and Hell's Half Acre lava fields.

Rhyolitic lava flows ooze slowly and move at most a few hundred feet a day, much slower than basaltic ones. These flows have covered areas ranging from a few to 250 square miles. The initial eruption of the largest ones covers areas around the vent several yards thick with ash and rock. Some of this could damage buildings, power lines and other structures, yet park staff should have enough time to safely evacuate areas, and once lava is flowing, they might even have enough time to move smaller structures out of harm's way. Yet lava flows would destroy everything in their paths that remain in place, including roads and buildings. They could also trigger wildfires in the summer months and dam or divert streams, causing flooding.

Again, scientists don't believe any volcanic activity—either a lava flow or a super eruption—will happen within thousands of years, but what does happen with more frequency are earthquakes and hydrothermal explosions.

People rarely feel the earthquake swarms in the caldera, which result from magma and hot water moving underground. The most powerful earthquakes in the Yellowstone region occur outside the caldera along active faults in the Gallatin, Madison and Teton Ranges.

About one to two large earthquakes rock Yellowstone Country each century. In 1959, not long before midnight one August night, the largest earthquake ever recorded in the Rocky Mountains shook Montana's Hebgen Lake area, a few miles west of the park. The quake triggered a landslide that started from a height of more than 1,000 feet, raced downslope, rushed through the Madison River and climbed 400 feet up the other side of the canyon. The dislodged rock and soil that settled at the bottom of the canyon dammed the river and created Earthquake Lake. Twenty-six people died that night, nearly all at the Rock Creek Campground, part of which the slide and surge of river water devastated. Two more died in a hospital several days later, and 19 of the dead remain entombed in the massive slide.

The Hebgen Lake earthquake, at magnitude 7.3, shook much of the West and supercharged Yellowstone's thermal areas. One hundred and sixty new geysers shot to life after the quake, some existing ones stopped erupting, and some dormant ones sprang back to life. And in Old Faithful's case, the time between eruptions grew longer.

Although Yellowstone's volcanism doesn't trigger major earthquakes, the park's hydrothermal explosions are linked directly to it. These explosions occur much more frequently than large earthquakes—about every two years—and are influenced by the heat radiating from the upper magma chamber.

Hydrothermal explosions forcefully shoot out steam, water, mud and rock with little to no warning. Smaller explosions eject rock several yards away while larger ones can erupt more than a mile into the air, with rock sometimes falling more than two miles from a crater. None of these explosions in Yellowstone have triggered eruptions of magma, which sits farther below the shallower hydrothermal systems. And smaller explosions occur much more often than larger ones.

By definition, hydrothermal systems need water and heat to power them. They also need earthquakes to keep the underground fractures open, which allows the superheated water to move through the system. When the plumbing of a hydrothermal system remains open, geysers and hot springs cycle the heated water to the surface. But when earthquakes or mineral deposits seal part of a system or when ground water declines seasonally, an explosion can occur when a swift drop in pressure causes water to boil suddenly and flash to steam, blasting apart overlying rock.

The Threat of the Supervolcano

The largest hydrothermal explosion known in the world formed the 1.5-mile-wide crater under Yellowstone Lake's Mary Bay about 14,000 years ago. It's one of 18 craters at least 300 feet wide formed since the last glaciation. Nearby Indian Pond is another crater from about 3,000 years ago and measures about 550 yards across.

More recently, smaller hydrothermal explosions in the 1880s and '90s in Midway Geyser Basin shattered the plumbing of Excelsior, once the world's largest geyser, turning it into a hot spring. In 1989, Porkchop Geyser exploded in the Norris Geyser Basin and shot out rocks at the feet of surprised onlookers. And in 2009, a group of scientists stood listening to a lecture at Biscuit Basin about hydrothermal explosions when a small one exploded nearby! The speaker happened to be Hank Heasler, a Yellowstone geologist, and the group was being led by Bob Smith, a professor from the University of Utah who has been studying Yellowstone's volcanism for nearly 50 years.

No one in any of the three witnessed explosions was scalded or injured by rock bombs, but the events underscore that these unpredictable incidents remain the greatest threat to people from the Yellowstone Volcano.

The Story of Yellowstone

Conclusion

The odds of the Yellowstone Volcano exploding anytime soon are extremely low, but another disaster with worldwide consequences has been steadily unfolding for decades. The damaging heat comes not from magma chambers or rushing lava but from the warmth trapped in Earth's atmosphere. As carbon dioxide and other greenhouse gases continue to be released by the burning of fossil fuels, more heat is trapped, and the planet continues to warm steadily. Vast stores of melting ice are raising sea levels that threaten coastal communities and ecosystems, and thunderstorms, hurricanes, flooding, wildfires, drought and heat waves are becoming more severe.

There is no refuge anywhere on Earth from the effects of climate change, including Yellowstone.

In the park and surrounding areas, winters are becoming shorter and milder, and summers are becoming longer, hotter and drier. At

Yellowstone's Northeast Entrance, at 7,350 feet in elevation, 80 more days on average stay above freezing each year than 50 years ago. Yellowstone also sees 30 fewer days with snow on the ground than in the 1960s. In Yellowstone Country in general, the area still receives about the same amount of annual precipitation it has for the past seven decades, but about 24 fewer inches accumulate as snow each year, falling as rain instead. And June and September are seeing less measurable snowfall overall.

Throughout the Yellowstone region, average daytime and nighttime temperatures continue to rise. The region has warmed 2.3 degrees Fahrenheit since 1950 and could double or even quadruple that rate by 2100. The shorter, warmer winters also result in thinner snowpack in the mountains; and warmer, wetter springs and hotter summers are melting that snow earlier. The annual streamflow peak is eight days earlier on average in the area's watersheds, and the runoff of snowmelt is shifting from June through August to March through May.

The most visible sign of a warming Yellowstone and western North America in general is the increasingly common wildfire smoke filling the summer skies. Fires have periodically burned in the Yellowstone area for more than 15,000 years, after the last glaciation. Although there have been warmer and drier periods with more frequent burns than today, the warming during those periods happened much slower—over thousands of years rather than decades—and the winter and nighttime temperatures were cooler then, too. Today, forests must adapt more quickly than in the past.

Because of warming, fire, drought and infestations, whitebark pines have already been pushed to the brink in much of Yellowstone and its ecosystem. Lodgepole pines, the most widespread tree by far in Yellowstone's forests and the most fire-adapted tree in the Rocky Mountains, are being challenged, too.

More acres burned in Yellowstone in 2016 than in any other year in the past century except for 1988. About half of the acres burned in 2016 engulfed young forests that had sprouted following the '88 fires. As the fire season continues to grow in length and as fires burn more frequently, lodgepoles will need to reach maturity and to produce cones to reseed themselves in less time than they've had to for thousands of years. When lodgepoles colonized the area 11,000 years ago, the climate was hotter and drier for about 4,000 years from that point on, and fires burned more often than they do today. What likely happened then and seems to be happening today is that Yellowstone's forests will become younger overall as fires become more common. Also, ponderosa pine, Gambel oak, western larch and other pioneering species could expand into Yellowstone as the ecosystem becomes more hospitable to them.

Some of Yellowstone's forests, however, might lack the time or moisture to regenerate after a fire, especially at elevations below 6,500 feet. In those places, sagebrush, juniper, grasses and other vegetation would likely expand. Non-native plants that are better adapted to warmer conditions and disturbances could also displace some of today's drought-intolerant species. Elk, bison, deer and pronghorn would have to adapt to changing food sources, and in many cases, invasive plants are unpalatable to them. Drier summers also yield less nutrition even from native plants.

Throughout all of Yellowstone's habitats, animals would need to adapt to the changing landscape, shift up in elevation, move to new areas, experience declining populations or die out. These animals include wolverines, which depend on deep snowpack to den and raise their kits; Canada lynx and snowshoe hares, the lynx's main prey, which could lose forested habitat; and pika, which are already declining in lower elevations across the West. Mammals and birds adapted to forests, including pine martens and boreal owls, could decline, as could songbird species that nest in Yellowstone's wetlands, many of which have been shrinking as the climate warms. Vanishing wetlands also affect other creatures, from frogs, salamanders and toads to sandhill cranes, trumpeter swans, beavers and moose.

In contrast, for the past several decades, the earlier melting snowpack has raised Yellowstone Lake and flooded the Molly Islands where American white pelicans, California gulls and double-crested cormorants have historically nested in large numbers. Caspian terns once nested there, too, but haven't since 2005. These waterfowl have also been affected by bald eagles preying on the young birds that do hatch successfully. The eagles' preferred food, cutthroat trout, has been decimated in the past three decades by illegally introduced lake trout, which prey on cutthroats and swim and spawn too deep for eagles to catch. Fewer cutthroats also means less food for pelicans and cormorants.

Also, with earlier runoffs, streams and rivers are running lower and warmer in summer. This affects trout and other fish and the birds and mammals that depend on them, including eagles, ospreys and otters.

People living throughout the West are also affected. The mountains in Yellowstone Country historically accumulated deep snowpack in winter, which gradually melted through late spring and summer and fed the headwaters of the Yellowstone, Snake and Green Rivers, which flow into the Missouri, Columbia and Colorado Rivers. With decreasing snowpack and earlier runoff, communities throughout the West that rely on the rivers for drinking water, irrigation, recreation and hydroelectric power are being affected. And the problem will only worsen the warmer the climate becomes.

The impacts of wildfire smoke on human health will also become increasingly problematic, as will the heat from hotter summers. The number of days above 90 degrees Fahrenheit are projected to increase each decade to as many as 60 to 90 by the end of the century in towns and cities in the Yellowstone area, including Idaho Falls,

Idaho; Bozeman and Livingston, Montana; and Cody, Jackson and Pinedale, Wyoming.

Humanity's impact on the climate is a significant force not unlike the glaciers and volcanos that have shaped Yellowstone for millions of years. To what degree the park and its ecosystem continue to change depends in large part on how quickly the world reduces its use of fossil fuels and slows the planet's warming. Yet that global issue also occurs alongside a local one that Yellowstone is facing.

The area's human population doubled in the past 50 years and could double again in the next 25. And the number of people visiting Yellowstone has also doubled since the early 1970s. With more people in the area, the limited supply of water will be further stressed, and private land development is fragmenting habitat in the valleys surrounding the park, in places that wildlife historically roamed and that some still depend on for winter forage. Also, fences and ongoing development block or impede ancient migration routes for elk, pronghorn and mule deer. And new and existing roads— with few underpasses and overpasses for wildlife to cross safely— take a toll on animals of all sizes. Other issues, including drought and more people recreating in remote areas, affect wildlife and the land, as well.

What Yellowstone might look like in the coming decades is uncertain. What is sure, however, is that it will continue to adapt as it has for millions of years—including after supervolcanic eruptions blew away mountains and glaciers scraped away everything to the bedrock.

The Story of Yellowstone

Acknowledgements

I am grateful to all the scientists, naturalists, writers and artists (past and present) whose works have made *The Story of Yellowstone* possible. They are too numerous to mention but I have benefitted greatly from their knowledge and inspiration.

Thanks also to those who provided feedback on an early draft of the book, including Kirsten Almberg, Mark Almberg, Ger Killeen, Amy O'Connell, Seán O'Connell, Tim O'Connell, Adrienne Papazian, Dan Phillips, Kate Saunders, Sarah Staggs, Rand Swanson and Gary Zylkuski.

And thanks to Courtney Oppel for copyediting and proofreading the final drafts of the book. Her edits and suggestions improved it considerably.

I am also indebted to Doug Henderson who allowed me to

include some of his evocative art in the book, including several unpublished works; to Ron Blakey for the use of his fascinating paleogeographic maps; and to Jacob W. Frank, Neal Herbert, Jim Peaco, Diane Renkin and others, whose excellent photography for the National Park Service is in the public domain and very much appreciated.

I am grateful, too, for the untold hours of camaraderie, wildlife watching and hiking with John Gann and Randi Mattson, whom I met early on in my trips to Yellowstone. They're some of the finest naturalists I've known, and their tirelessness, enthusiasm and curiosity are contagious.

Thanks also to the many other people I've shared time with in the park and adjoining lands while watching wildlife, hiking, skiing and enjoying other activities. I am especially appreciative of the dozens of people I have backpacked with. The most steadfast of these has been Ger Killeen. From the Gallatin Skyline to the Thorofare and everywhere in-between, I couldn't ask for a better trail and fireside companion.

Special thanks, as well, to my brother Seán for introducing me to Yellowstone, for our time in the park since then—including my first attempt at backpacking (a grizzly in camp turned us back)—and for reviewing the book during its different stages.

Lastly, thanks to Emily and Maeve for their love and support and for our adventures in Yellowstone and beyond.

The Story of Yellowstone

Kickstarter Supporters

The Story of Yellowstone began as a successful crowdfunding project on Kickstarter. Thanks to the following for their support and patience while I researched and wrote the book:

Sophia Acord

Pete Alaimo

Emily Almberg

Mark Almberg

Camellia El Antably

Susan Apstein

Albert Archer

Laura Baughman

Lorelle Berkeley

Luke Blosser

Eric Borchert

Angela Brennan

Scott Brennan

Bob Brill

Chris Brown

Nate Bunnyfield

J. Campbell

Kent Campbell

Michael Campbell

Mark Cavanagh

The Story of Yellowstone

Marla Clancy

Lauren Coleman

Doug Colson

Christy Connolly

John Conwell

Lonna Cosmano

Andrew Crowley

Kim Curren

Samantha Curren

Christopher Davis

Aroldo de Rienzo

Mark Delwiche

Jesse DeVoe

Nancy Dobson

Paul Dolan

Paul Driver

Kevin Dunn

Patricia A. Durkin

Jared Duval

John Fiscus

Sheila Fiscus

Monica O'C Fleming

Chris Forristal

Vicki Forristal

Ken Foszcz

John Gann

Tomas Gradin

Bob Himes

Mike Himes

Sr. Mary Andrew Himes

Peter Hudson

Chih-Ming Hung

Mary Ann Jenkins

Jenny Jones

Pauline Kamath

Bess Katerinsky

Joanne Miller Keefe

Mary Ann and Wayne Kendall

Robert Khoe

Ger Killeen

Kickstarter Supporters

Paul Klein

Paul Kukkonen

Walter and Fran Kukkonen

Matthew T. Lavin

Amber D. Luthi

Katharine Macaulay

Elizabeth Mathur

Michael McCaffrey

Rick McIntyre

Christy Hale McKenna

Jerod Merkle

Matt Metz

Jim Miani

Sharon Nissen

Daniel Nissman

Rolf Nitsche

Amy O'Connell

Dorothy O'Connell

Seán O'Connell

Tim O'Connell

Lisa Ohlinger

Sonja Oliveri

Carl Owenby

Alex Ozz

Andy Phelan

Dan Phillips

Tony Polecastro

Jackson R. Pope III

Nancy Pope

Joanna Rayburg

Eduardo Santiago

Kate Saunders

Francisco Scaramuzza

Suzanne Sharp

JC Shei

Nichole Simmons

Erna Smeets

Becky Smith

Scott Snow

Jeanne Stacey

The Story of Yellowstone

Jane Staggs

John Staggs

Sarah Staggs

Sue Ellen Staggs

Tom Staggs

Brandan Still

Sadako Tengan

Julie Timmons

James Turnbull

Nathan Varley

Samuel and Nadine Vlasta

Nancy Abbatiello Warren

Anne Whitbeck

Donna Wilson

Katie Yale

Linda Yates

Andrew Yen

Gary Zylkuski

The Unknown

Bibliography

Beschta, Robert L., and William J. Ripple. 2019. "Large Carnivore Extirpation Linked to Loss of Overstory Aspen in Yellowstone." *Food Webs* 22, e00140 (December): 1-7.

Bonner, Hannah. 2015. *When Fish Got Feet, When Bugs Were Big, and When Dinos Dawned: A Cartoon Prehistory of Life on Earth.* National Geographic Society.

Breining, Greg. 2007. *Super Volcano: The Ticking Time Bomb Beneath Yellowstone National Park.* Stillwater, OK: Voyageur Press.

Brock, Thomas D. 1994. *Life at High Temperatures.* Yellowstone Forever.

Christiansen, Katie Shepherd, ed. 2021. *The Artist's Field Guide to Yellowstone: A Natural History by Greater Yellowstone's Artists and Writers.* San Antonio, TX: Trinity University Press.

Christiansen, Robert L. 2001. *The Quaternary and Pliocene Yellowstone Plateau Volcanic Field of Wyoming, Idaho, and*

Montana. US Geological Survey.

Christiansen, Robert L., Jacob B. Lowenstern, Robert B. Smith, Henry Heasler, Lisa A. Morgan, Manuel Nathenson, Larry G. Mastin, L. J. Patrick Muffler, and Joel E. Robinson. 2007. *Preliminary Assessment of Volcanic and Hydrothermal Hazards in Yellowstone National Park and Vicinity*. US Geological Survey.

Cornwall, Warren. "Have Returning Wolves Really Saved Yellowstone?: Researchers Fear That Some Damage Can't Be Undone." *High Country News*, Dec. 8, 2014.

Craighead, Charles, and Henry H. Holdsworth. 2006. *Geology of Grand Teton National Park*. Grand Teton Association.

DK. 2012. *Prehistoric Life: The Definitive Visual History of Life on Earth*. New York, NY: Dorling Kindersley.

Fishbein, Seymour L. 1997. *National Geographic Park Profiles: Yellowstone Country*. National Geographic Society.

Flannery, Timothy. 2002. *The Eternal Frontier: An Ecological History of North America and Its Peoples*. London, England: Vintage.

Fouke, Bruce W., and Tom Murphy. 2016. *The Art of Yellowstone Science—Mammoth Hot Springs as a Window on the Universe*. Crystal Creek Press.

Franke, Mary Ann. 2000. *Yellowstone in the Afterglow: Lessons from the Fires*. National Park Service.

Bibliography

Frasier, Rhonda. 2022. *Bison History Timeline*. https://allaboutbison.com/bison-in-history/bison-timeline.

Fritz, William J., and Robert C. Thomas. 2011. *Roadside Geology of Yellowstone Country*. Mountain Press.

Good, John M., and Kenneth L. Pierce. 2016. *Interpreting the Landscape: Recent and Ongoing Geology of Grand Teton & Yellowstone National Parks*. Grand Teton Association.

Grant, Richard. 2021. "The Lost History of Yellowstone: Debunking the Myth that the Great National Park Was a Wilderness Untouched by Humans." *Smithsonian Magazine*, January 2021.

Handwerk, Brian. 2022. "Five Big Changes Scientists Have Documented During Yellowstone National Park's 150-Year History." *Smithsonian Magazine*, April 7, 2022.

Hazen, Robert M. 2013. *The Story of Earth: The First 4.5 Billion Years: From Stardust to Living Planet*. Penguin Books.

Hendrix, Marc S. 2011. *Geology Underfoot in Yellowstone Country*. Mountain Press.

Holloway, Marguerite. "Your Children's Yellowstone Will Be Radically Different." *New York Times*, Nov. 15, 2018.

Horner, Jack. 2004. *Dinosaurs under the Big Sky*. Mountain Press.

Hostetler, Steven, Cathy Whitlock, Bryan Shuman, David Liefert, Charles W. Drimal, and Scott Bischke. 2021. *Greater Yellowstone*

Climate Assessment: Past, Present, and Future Climate Change in Greater Yellowstone Watersheds. Bozeman, MT: Montana State University, Institute on Ecosystems.

Huang, Hsin-Hua, Fan-Chi Lin, Brandon Schmandt, Jamie Farrell, Robert B. Smith, and Victor C. Tsai. 2015. "The Yellowstone Magmatic System from the Mantle Plume to the Upper Crust." *Science* 348, no. 6236 (April): 773–776.

Johnson, Jerry, ed. 2010. *Knowing Yellowstone: Science in America's First National Park.* Lanham, MD: Taylor Trade Publishing.

Keefer, William R. 1971. *The Geologic Story of Yellowstone National Park.* US Geological Survey.

Kenedi, Christopher A., Steven R. Brantley, James W. Hendley II, and Peter H. Stauffer. 2004. *Volcanic Ash Fall—A "Hard Rain" of Abrasive Particles.* US Geological Survey.

Klaptosky, John. 2016. "The Plight of Aspen: Emerging as a Beneficiary of Wolf Restoration on Yellowstone's Northern Range." *Yellowstone Science* 24, 1: 65-69.

Lange, Ian. 2002. *Ice Age Mammals of North America: A Guide to the Big, the Hairy, and the Bizarre.* Mountain Press.

Lahren, Larry. 2006. *An Archaeologist's View of Yellowstone Country's Past.* Cayuse Press.

Love, J. David, John C. Reed Jr, and Kenneth L. Pierce. 2016. *Creation of the Teton Landscape.* Grand Teton Association.

Bibliography

Lowenstern, Jacob B., Robert L. Christiansen, Robert B. Smith, Lisa A. Morgan, and Henry Heasler. *Steam Explosions, Earthquakes, and Volcanic Eruptions—What's in Yellowstone's Future?* US Geological Survey, 2005.

MacDonald, Douglas H. 2018. *Before Yellowstone: Native American Archaeology in the National Park.* University of Washington Press.

MacNulty, Daniel R., Daniel R. Stahler, C. Travis Wyman, Joel Ruprecht, and Douglas W. Smith. 2016. "The Challenge of Understanding Northern Yellowstone Elk Dynamics after Wolf Reintroduction." *Yellowstone Science* 24, 1: 25-33.

Marcus, W. Andrew, James E. Meacham, Ann W. Rodman, Alethea Y. Steingisser, and Justin T. Menke. 2022. *Atlas of Yellowstone: Second Edition.* Berkeley, CA: University of California Press.

McEneaney, Terry. 1989. *Birds of Yellowstone: A Practical Habitat Guide to the Birds of Yellowstone National Park—And Where to Find Them.* Roberts Rinehart.

Metz, Matthew C., Douglas W. Smith, Daniel R. Stahler, John A. Vucetich, and Rolf O. Peterson. 2016. "Temporal Variation in Wolf Predation Dynamics in Yellowstone: Lessons Learned from Two Decades of Research." *Yellowstone Science* 24, 1: 55–60.

Montana State University Thermal Biology Institute, and the Montana Institute on Ecosystems. 2013. *Living Colors: Microbes of Yellowstone National Park.* Yellowstone Association.

Nabakov, Peter, and Lawrence Loendorf. 2004. *Restoring a Presence: American Indians and the Yellowstone National Park*. Norman, OK: University of Oklahoma Press.

Peterson, Christine. 2020. "25 Years after Returning to Yellowstone, Wolves Have Helped Stabilize the Ecosystem." *National Geographic*, July 10, 2020.

Phillips, Michael K., and Douglas W. Smith. 1998. *The Wolves of Yellowstone*. Voyageur Press.

Picton, Harold D., and Terry N. Lonner. 2008. *Montana's Wildlife Legacy: Decimation to Restoration*. Media Works Publishing.

Pim, Keiron. 2014. *Dinosaurs—The Grand Tour: Everything Worth Knowing about Dinosaurs from Aardonyx to Zuniceratops*. The Experiment.

Prothero, Donald R. 2016. *The Princeton Field Guide to Prehistoric Mammals*. Princeton, NJ: Princeton University Press.

Quammen, David. 2016. *Yellowstone: A Journey through America's Wild Heart*. National Geographic Society.

Quammen, David. "Yellowstone: The Battle for the American West." *National Geographic*, May 2016.

Ray, Andrew M., et. al. 2019. *Vital Signs—Monitoring Yellowstone's Ecosystem Health*. National Park Service.

Ripple, William J. and Robert L. Beschta. 2012. "Trophic Cascades in

Yellowstone: The First 15 Years after Wolf Reintroduction." *Biological Conservation* 145, 1: 205–213.

Ruth, Toni K., Polly C. Buotte, and Maurice G. Hornocker. 2019. *Yellowstone Cougars: Ecology before and during Wolf Restoration.* University Press of Colorado.

Schullery, Paul. 2004. *Searching for Yellowstone: Ecology and Wonder in the Last Wilderness.* Montana Historical Society Press.

Schullery, Paul, and Lee H. Whittlesey. 1995. "Summary of the Documentary Record of Wolves and Other Wildlife Species in the Yellowstone National Park Area Prior to 1882." In *Ecology and Conservation of Wolves in a Changing World*, edited by Ludwig N. Carbyn, Steven H. Fritts, and Dale R. Seip, 63–76. University of Alberta Press.

Schullery, Paul, and Lee H. Whittlesey. 1999. *Early Wildlife History of the Greater Yellowstone Ecosystem. An Interim Report Presented to the National Research Council, National Academy of Sciences Committee on Ungulate Management in Yellowstone National Park.* National Park Service.

Smith, Robert B., and Lee J. Siegel. 2000. *Windows into the Earth: The Geologic Story of Yellowstone and Grand Teton National Parks.* New York, NY: Oxford University Press.

Steelquist, Robert. 2002. *Field Guide to the North American Bison.* Seattle, WA: Sasquatch Books.

Travis, Lauri. 2019. *Arrowheads, Spears, and Buffalo Jumps: Prehistoric Hunter-Gatherers of the Great Plains.* Mountain Press.

Wallace, David Rains. 2001. *Yellowstone: Official National Park Handbook*. National Park Service.

Wilmers, Christopher C., Matthew C. Metz, Daniel R. Stahler, Michel T. Kohl, Chris Geremia, and Douglas W. Smith. 2020. "How Climate Impacts the Composition of Wolf-Killed Elk in Northern Yellowstone National Park." *Journal of Animal Ecology* 89, 6 (June): 1511–1519.

Yellowstone National Park. 2021. *Yellowstone Resources and Issues Handbook: 2021*. National Park Service.

Author's note: While researching *The Story of Yellowstone*, I turned to countless web sites for information. The sites I visited most frequently include the ones for: the National Park Service, the US Geological Survey (including the Yellowstone Volcano Observatory), NASA, the BBC, National Geographic, PBS (including NOVA), NPR, EarthSky, the American Museum of Natural History, Smithsonian Magazine, the Bozeman Daily Chronicle, the Billings Gazette and Wikipedia. I also made many trips to the Museum of the Rockies in Bozeman, Montana, to learn about the area's geology and prehistory.

Index

Index

Index

Index

Catching Breakfast **by Maeve O'Connell**